BIO-BASED PLASTICS: MATERIALS AND APPLICATIONS

BIO-BASED PLASTICS: MATERIALS AND APPLICATIONS

Edited by:

Navodita Bhatnagar

ARCLER
PRESS

www.arclerpress.com

Bio-Based Plastics: Materials and Applications

Navodita Bhatnagar

Arcler Press
2010 Winston Park Drive,
2nd Floor
Oakville, ON L6H 5R7
Canada
www.arclerpress.com
Tel: 001-289-291-7705
001-905-616-2116
Fax: 001-289-291-7601
Email: orders@arclereducation.com

ISBN: 978-1-77407-277-6 (Hardcover)

ABOUT THE EDITOR

Navodita Bhatnagar finished her masters in technology and currently pursuing her PhD in environmental sciences from Institute of Technology, Carlow, Ireland. She's a keen enthusiast of environment protection and management. An alumunus of Indo-German centre for sustainibility which is essentially a collaborative interface to gather scientists in the field of environment sciences to come together and work for the common cause which is environment protection.

TABLE OF CONTENTS

LIST OF FIGURES

LIST OF TABLES

PREFACE

Plastics are an important part of the world in the current times with an extremely wide range of applications in several industries and places across the globe. The recent concerns over their harmful impact on the environment have given rise to the research and production of bioplastics, which causes a negligible impact on the environment. This book discusses some of these bioplastics which are prominent in the industrial setup in the modern world.

Some salient features of this book are:

- The book introduces the topic of bioplastics to the readers by starting with the definitions and concepts. It moves on to explain the different classifications of bioplastics that are in use and being manufactured in the current times.
- The book goes on to discuss the history of bioplastics and their evolution. It discusses the market potential for the bioplastics and their possible demands in the future. It talks about the biosynthesis of bioplastics along with other methods of production. It also discusses the biodegradable plastics that may be manufactured by the organisms that are genetically modified.
- The book emphasizes the importance of replacing the conventional forms of plastics in the market and their market share. It shares the impact that bioplastics may have on the recycling processes that are currently being undertaken. It also focuses on the different kinds of environmental impacts of the bioplastics and compares them with those of the conventional plastics and their products.
- The book emphasizes the importance of bioplastics in the future and its impact on the future economy in a bio-economy and the role of bioplastics in the future refineries.
- Moving forward, the book brings the attention of the readers to the starch -based bioplastics and its importance in renewable and biodegradable materials. It emphasizes on the capability of starch -based bioplastics to replace various petroleum-based products which may help in reducing the carbon footprint of those products.
- The book informs the readers about the different kinds of starch-

based plastics and how bioplastics can be manufactured from the food crops that contain starch. It explains the structure and properties of the starch-based plastics. It discusses various possible applications of the biodegradable plastics that are made out of starch and also focuses on the future applications of starch in some packaging materials.

- Then, the book talks about the cellulose compound that can be found in wood and its importance in the formation of bioplastics. The book throws light on the structure of cellulose and its sources. It also brings the attention of the readers to the cellulose fibers and their chemical composition.
- The book discusses the degradation process of cellulose based products and informs the readers about the time they take to degrade. It also gives insights on the materials that can be derived from cellulose acetate and the end products that they may be converted to. It discusses the research that is happening in the direction of cellulose-based products and the future prospects of such research.
- Going further, the book talks about another raw material for the production of bioplastics and bio-based materials which is chitin and its derived form, chitosan. The book discusses the sources in which chitin can be found and highlights the various uses of chitin.
- It explains the characteristics of chitin to the readers for better understanding of its uses and the structure of both these compounds.
- It also informs the readers about the synthesized form of chitin, that is chitosan. It discusses the various industrial applications of chitosan and its importance in industries like water processing, paper industry and so forth.
- The book discusses AROBOFORM and its source, lignin. The book describes its structure and the places where it can be found. It emphasizes on the various uses of lignin and talks about its importance in the future of natural products. The book lays down the various innovative aspects of ARBOFORM including social, economic and environmental ones.
- Moving forward, the book informs the readers about one of the largest sources of bioplastics and the most commonly used material – Polylacticacid (PLA). It dwells upon the process of crystallization that enables the appropriate usage of PLA and enlists the various industrial applications of it. It also shows and explains the structure, composition and production process of PLA.

- The book explains the readers about the meaning of sustainability in terms of bioplastics. It informs them about the principles that should be considered in the production of bioplastics. It talks about the importance of measuring the carbon footprint of various materials and the life cycle assessment, discussing the future of the sustainable form of plastics.

Why this book?

Plastics are being used across the world in high volumes and in day to day applications. Their concentration levels are increasing in a very harmful manner as they have been found to degrade very slowly which has a negative impact on the environment.

Chapter 1

INTRODUCTION TO BIO-BASED PLASTIC: MATERIALS AND APPLICATIONS

CONTENTS

In the past ten years, the field of bio-based plastics has significantly developed. The term "bio-based" plastics usually refers to the plastic which is of bio-based origin. The term "Bio-based" is not synonymous to the term "biodegradable". Currently, bio-based plastics represents the latest overview of the fundamental aspects of bioplastics and their applications. The chapter basically deals with the definitions and history of the bio-based plastics, market and their environmental impact. At the end of this chapter, future trends have been briefly described focusing on the materials that are in use currently.

1.1 INTRODUCING BIO-BASED PLASTICS

1.1.1 Various Definitions of Bioplastics

There is no one definition of sustainability. A product can possess various characteristics of sustainability. These characteristics can be either related or independent of each other and help in determining the sustainability of a product. Here are some definitions of bioplastics:

- On the basis of properties, a material can be defined as bioplastics if it is biodegradable, bio-based or includes both the properties. Bioplastics are those materials which possess a number of applications and properties.
- Bio-based materials are obtained from organic matter that consists of a whole or just a part of biological carbon that can be used as a replacement of fossil. Bio or renewable carbons are preferred in place of conventional fossil fuels such as petrol, diesel, kerosene etc. Materials are certified as bio-based by the USDA Bio preferred Program in the United States. Items made using bio-based materials that are certified can see favored purchase by Federal procurement agencies.
- Traditional plastics are combined with biomaterials for instance flax, hemp, wood, jute, starch and other similar materials to form bio composite material. Such materials not only reduce the proportion of petroleum-based non-renewable plastic but are also used to enrich physical characteristics and infuse a natural aesthetic.
- Biodegradation can be defined as the process of converting materials into natural substances like carbon dioxide, water and

compost with the help of microorganisms that are present in the environment. Biodegradable materials are those materials which can be degraded by the action of microorganisms. This chemical process of biodegradation depends on several factors such as environmental conditions like weather, temperature, etc.

- ASTM International standards 6400 and 6868 define products which are compostable plastics but there are certain requirements of these standards. The requirements for international standards include products which biodegrades in a certain time period without leaving any toxic compounds in the soil.

Renewable feedstock

- In plastic production, the amount of harmful gaseous compounds that majorly include greenhouse gases can be substantially reduced by using organic materials such as natural fibers and starch. These organic materials act as a good substitute for petroleum-based feedstock. In the manufacturing phase, bio-based plastic resins behave in a similar manner as the traditional plastics. These organic bio-based plastics also keep most of the feel and look of the plastics that are petroleum based.

Reclaimed feedstock

- A byproduct of one industry is a way to sustainability for another industry. Petroleum-based feedstock used to make traditional plastics can be substituted by using various waste materials like agricultural waste products and wood fibers from milling operations. Taking into consideration the potential of reclaimed feedstock, it can replace nearly 70 percent of non-renewable resources that are used in the plastic industry, globally. With a boom in the market of reclaimed feedstock, it could become a serious concern for the quantity of fossil fuels utilized in the plastic industry.

Biodegradable materials

- A huge reduction in landfill waste could be seen by the use of compostable plastics, especially when utilized in food services along with the compostable food waste and in various packaging applications. Biodegradability has a great functional advantage especially for some specific industries like products related to mulch films.

Recycled and upcycled goods

- Massive savings in energy consumptions can be obtained just by opting recycled plastics over virgin plastics. Recycled materials and upcycling of reclaimed plastics reduce the environmental footprint when these materials are used in the form of biocomposites. Using them for high-quality durable goods extensively reduces the impact on nature. These recycled and upcycled goods can expand and create a market for waste materials which otherwise would be put in landfills.

Classification of Bioplastics

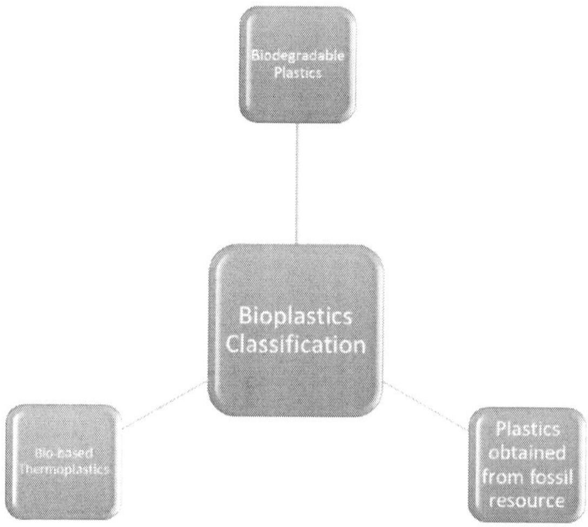

Figure 1.1: Classification of bioplastics.

Bioplastics can be classified into three major groups:

- Bio-based and biodegradable plastics, such as starch-based blends (TPS), potentially polybutylene succinate (PBS), polylactic acid or polylactide (PLA), cellulose acetates and polyhydroxy alkanoates (PHA's).
- Bio-based thermoplastics or partly bio-based non-biodegradable thermoplastics, for example, bio-based technical performance polymers like nylon 11 (based on amino undecanoic acid from castor oil), polytrimethylene terephthalate (PTT) and other polyamides (PA) and bio-based PP, PE, PVC or PET.

Bio-based thermosets are also included in this category. Few examples of bio-based thermosets are soy-based thermosets, epoxies and unsaturated polyesters.

- Plastics which are biodegradable and are presently fossil resource based.

Polybutylene succinate (PBS), Polycaprolactone and Polybutylene adipate terephthalate (PBAT) are the most noteworthy plastics of these. Over 95% of the total action in bioplastics is of these specific materials. They can also be used to discuss key characteristics, trends in the industry and projected eye-catchy market segments OD dislocation.

1.1.2 A Brief History of Bio-based Plastics

"Steadily growing since 1950, the worldwide production of plastics materials has reached 288 million tons in 2012" (Plastics Europe, 2013). Germany is one of the most important actors in the plastic markets and the biggest manufacturer and processor of plastics in the EU.

"In 2012, the German plastic industry produced 19.5 million tons of plastics and generated a revenue of 25.1 billion euros" (Plastics Europe Deutschland, 2013). "Large amounts of petroleum, approximately 7% of the global production, are used for the production of plastics" (Bozell, 2008).

Subsequently, the debate on today's (renewable) energy is also affecting the plastic industry. The challenges are well known and essentially the same that underlie the energy debate. There is limited availability of fossil resources like gas and petroleum. Also, they are volatile in nature and their prices are increasing day by day. As there is an increase in the utilization of plastics more pressure is created on the decision makers by the growing political and public concern about the harmful effects of it on the environment.

The biggest challenge to the plastic industry is to meet the energy requirements, as the fossil resources are exhausting. This challenge is bigger than it might seem. The energy issue has only one alternative i.e. the use of biomass. This alternative is renewable energy to produce bulk chemicals and plastics.

The manufacturing of plastics by the method that involves use of biomass is no new science. This approach was used to develop plastics since the beginning of the last century, for example, production of celluloid which is the first modern artificial polymer and the process of vulcanization of rubber are quite established processes.

However, during the petrochemical revolution, the petroleum-based plastics, which we are using today, developed a new market in place of the plastics which are based on renewable resources.

Though the functional and physical variety of plastic seems giant, there are five commodity plastics which account for 64% of the processed volume. These five commodity plastics are:

- polypropylene (PP)
- poly (ethylene terephthalate) (PET)
- polyethylene (PE)
- poly (vinyl chloride) (PVC)
- poly(styrene) (PS)

Engineering plastics like polyamides and polyurethanes and others account for the remaining 36% volume.

The amount of plastics that is processed has applications in many sectors. The major share of this is used in construction and packaging sector, followed by electronics and automotive applications. Plastics offer many social, economic and ecological benefits and have become a vital part of the present lifestyle.

Few examples are handy packaging that limits food spoilage, very proficient insulation foams that helps us in building zero-energy buildings and lightweight plastic components that reduce the consumption of fuel in vehicles and aircraft. The plastics we are using in the present day have only one drawback that they cannot be renewed. The use of the bio-based feedstocks as the raw material base signifies an appealing way out for many in the plastic industry.

Manufacturing of bio-based plastics was an objective of many industries in the past three decades. The related research action shows that "bioplastics" is a diverse term and cannot be defined by general characteristics. To be precise, here is the basic definition: Any plastic material that is either biodegradable, bio-based (completely or partially) or possesses both the properties is termed as bioplastic.

The main focus lies on deriving the bio-based plastics necessarily from renewable resources that may or may not be biodegradable. The foremost bio-based product presented to the market were poly (hydroxyalkanoate), which was manufactured by ICI (Great Britain) in the 1980s. The development of this product increased with the increase in its demand by the agricultural and packaging sector.

These important products could never make it through the important markets because of two major drawbacks namely, their high prices and deficiency of important properties needed for chief applications. In spite of these drawbacks, high interest remained in biodegradable products. It was because the number of applications increased beyond the originally very limited number of simple packaging applications.

In 1990s biodegradable products were developed using materials with improved properties. Consumers started getting irritated by these biodegradable products as some of them were completely derived from petroleum.

Towards the end of the 1990s, more interest grew in the use of renewable raw materials than in bioplastics. "In 2001, Nature Works LLC was the first company that produced a completely bio-based plastic (poly (lactic acid)) in large quantities" (USA, 140ktpa).

Since then, the manufacturing of various bio-based plastics has been demonstrated at pilot and demo scale. Some of the products are completely new plastics, for instance, poly (lactic acid), some are partially bio-based like polyols, polyamides and some are conventional plastics, for example, PVC (polyvinyl chloride) and PE (polyethylene).

To make the plastics at more reasonable prices and to increase their scope, petrochemical monomers are often required in small amounts for bio-based plastics whose replacements cannot be resembled.

With recent development in the field of bio-based plastics along with improvised technology and research, about ninety percent of plastics that is used by us can easily be substituted (Shen et al, 2010). One of the main reasons that are preventing the greater reach and deeper market reach is the high pricing of bioplastics and limited production capacities if bio-based plastics.

In regards with a recent forecast study performed by European Bioplastics, Institute for Bioplastics and Bio composites in 2013 published in Plastics Europe, bioplastics can only be produced in very small amounts that approximately correspond to 0.5 percent of global production of plastics.

Though these initial figures are not that attractive and small in number, bio-based plastics have shown exponential growth and a bright future in the field of commerce and in ongoing commercialization.

The plastic industry is rapidly progressing towards bio-based products. If the bio-based plastics keep increasing at the same pace as they are, they will definitely play an important role in the forthcoming rise of bio-economy.

1.2 RECOVERY AND PRODUCTION OF BIOPLASTICS

From renewable sources, the bioplastics can be recovered easily with the help of microorganisms. These are the polyesters that are extensively spread all over nature and can be accumulated in the intracellular arrangement in microorganisms that are stored in form of granules that possesses the physio chemical properties that mimic the exact same properties as of petrochemical plastics.

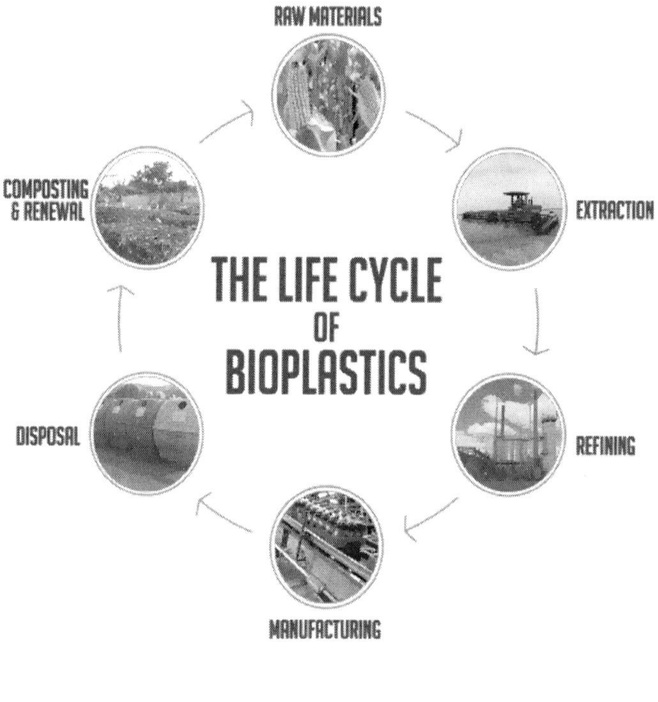

Figure 1.2: Life-cycle of bio-based plastics.

Source: http://www.thegreeneconomy.com/content/shifting-towards-bio-based-economy

These days, technology in the field of bio plastics is touching heights, enabling the production of bioplastics from many plant sources and also from micro-algae. Algae produce a number of base materials that can be used in different steps during bio-plastics production.

Most important by-products obtained are hydrocarbons and carbohydrates. For example, taking the case of algae *Botryococcus braunii*, that has the capacity to produce and excrete these materials into another medium.

As per the Annual Market data update, it can be estimated that the capacity of bio-based plastics has increased more than four times from around 1.6 million tonnes in the year 2013 to around 6.7 tonnes in the year 2018.

The process of Bioplastics production does not impact bio-degradability of the product. The procedures for bio-plastics production can be

- biotechnological or synthetic
- the most popular method that includes preparation of plastics with the help of natural polymer, that is chemically or mechanically treated
- chemical synthesis of a monomer-based polymer, acquired by biotechnologically converting renewable resources
- using lactic acid obtained from the process of fermentation of Sugars to produce polylactic acid (PLA)

There is a massive reduction in carbon dioxide emissions in the production of most bioplastics as compared to conventional and traditional techniques. But as every coin has two sides, there are rising concerns over the production of bio-plastics as well that the creation of a global bio-economy will need to produce bioplastics in huge amount and that could chip into an increased rate of soil erosion and deforestation, directly impacting the water supplies adversely.

Glucose is utilized in place of fructose by Wild-type strains of *R. eutropha*. The strains that consume glucose have been specifically developed to be particularly used in the production of BIOPOL on a commercial scale. Bacteria that produce PHAs can be grown on an industrial level in large quantities up to high cell densities in a stirred tank with the help of sucrose, molasses or glucose as a carbon source.

To increase the efficiency and yield, fed-batch systems are preferred over continuous processes and a rise of around 70 to 80 percent in the output of PHA has been recorded for *R.eutropha* grown in a mineral salts medium supplemented with glucose taking the single carbon source in form of propionate. The cell densities were reported to be in the order of 100g dry weight lifter 1. By undertaking cell harvesting by flocculation or by centrifugation, the PHA can be obtained with the help of hypochlorite and surfactants to lyse the cells and discharge the intracellular product.

Though hypochlorite can facilitate an appropriate laboratory procedure, it can also be a reason for some degradation of the product. PHA can be separated from soluble intracellular products by floatation processes. Solvent extraction has been proposed as an alternative.

In order to achieve this effectively, the cell mass should first be dried either using lyophilization or by spray drying. This procedure also adds to the cost. Huge quantities of solvent are required as concentrated solutions (that are almost viscous) of PHA. Most of the PHA products can be dissolved by methylene chloride or chloroform but it is needed to treat the cells before in order to achieve extraction with maximum efficiency. The actual yield of the product and its composition is completely dependent on the chosen substrate regime.

Choi and Lee observed the cells with low polymer content because the residue propionic acid content was high in the medium. If the feed strategy was altered, by using oleic acid addition and acetic acid induction, P (3HB-co-3HV)-polymer yields of 78 percent wt. and productivity of 2.88 g l-1 h-1 was obtained.

1.2.1 Occurrence and Composition

In prokaryotic cells, polyhydroxy alkanoic acids occur as storage polymers which are presently well-known. These compounds have a general structure and are soluble in water. poly-β-hydroxybutyric acid (PHB), an intracellular carbon and energy storage compound is produced by many bacteria in large quantities.

This property though absent from enteric species, but is extensively found in Pseudomonads and related species which include Rhizobium bacteria which shares a symbiotic relationship with leguminous plants like beans, lentils, peas etc. and nitrogen-fixing bacteria, *Azotobacter* spp.

In the existence of an unlimited supply of energy and carbon source, unbalance growth occurs which results in accumulation. Under favorable conditions, the polymer has a tendency to aggregate to more than 50-80 percent of the weight of the dry cell.

In the cytoplasm, the product is stored as granular inclusion bodies. However, these compounds are not suitable as potential bioplastics because their chain length is short. and. Archaea including *Haloferax mediterranei* are species which synthesize PHV and PHB. These halophilic archaea might come as an advantage for production because their culture requirements of salinity and comparatively, high temperature facilitates the small opportunity for the development of contaminants.

The recovery of PHB becomes difficult when the species like *Azotobacter vinelndii* produces huge amounts of exopolysaccharides which diverts the substrate to alternative products. Production of mutant strains which give high yields resulted in conversion rates of 65 percent for PHB and eventual PHA yields of 71 percent dry weight.

1.2.2 Biosynthesis

Due to the involvement of only three enzymes, namely acetoacetyl CoA reductase, PHB synthase and β-thioketolase, the biosynthesis of PHB is simpler in comparison to the formation of most polysaccharides. The genes involved are found in *Pseudomonas oleovorans* and *Ralstonia eutropha* (earlier called *Alcaligenes eutrophus*) as well as many other species.

The genetic control involved in the process is quite simple. It involves three genes that are organized in an operon to form an array of three open reading frames. Also, these can be transmitted more easily to other species of bacteria and to yield plant species which are transgenic.

A revised alternative pathway was defined by Steinbüchel and Füchtenbusch which was found in *Rhodospirillum rubrum*. On the basis of difference in the constitutive protein and their substrate specificity, the gene products of PHA synthase from different bacteria were grouped into three different categories. Current research using recombinant strains of *E. coli* in which genes of PHA biosynthesis from *R. eutropha* has been inserted gave a yield of PHB with a mass of 3-11 x 10^6 Da.

From this product, the mechanical properties of this product were enriched by extending over 400 percent. The enzymes are generally constitutive for the synthesis of PHB and regulation of synthesis is seen at the enzyme level.

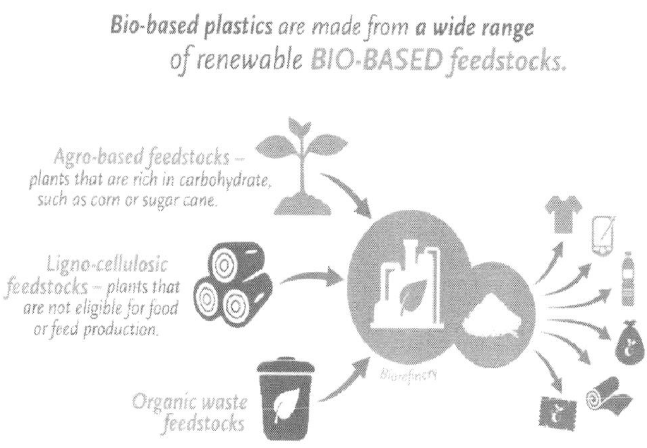

Figure 1.3: Bio-based feedstocks to produce bio-based plastics.

Source: https://www.european-bioplastics.org/how-much-land-do-we-really-need-to-produce-bio-based-plastics/

1.2.3 Products

Bioplastics differ from the products of the petrochemical industry as they are thermoplastic compounds which are biodegradable. Further, they also have an advantage that they can be produced using renewable resources. They are optically active, generally quite crystalline and possess piezoelectric properties.

A considerable amount of effort was put in by ICI (Zeneca) in the development of bioplastics and soon discovered that PHB did not possess all the characteristics that they wanted.

However, they found that it could manufacture linear co-polymers constituting hydroxyvaleric acid and poly-β-hydroxybutyric acid. These types of random copolymers were synthesized using glucose with *R. eutropha* and propionic acid as substrates. Typically, polymers are composed of 12-20 percent hydroxyvalerate and 300000 Da. approximately.

The melting point of the co-polymer is inversely proportional to the hydroxyvaleric acid content. This means, as the hydroxyvaleric acid content will increase, the melting point of the copolymers will decrease. The PHB has a melting point nearly around 175^0C while the melting point of PHAs is comparatively lower.

The PHBs are brittle whereas the copolymers are elastic. PHA are thermoplastic polymers. They can be molded because of their property of turning quite viscous when the temperature crosses their melting point.

The product's composition decides three things, namely:

- Crystallinity
- Melting point (T_m)
- Transition temperature (T_g)

Many products were formed by the commercialized product 'Biopol', for example, disposable razors, golf tees, biodegradable plastic bottles, and other products.

However, this production declined because of the higher cost of the product as compared to the plastics formed by chemical synthesis. Many alternative applications have been proposed, like water impermeable coatings for biodegradable packaging, as temporary plates and pegs in the repair of bone injuries etc. but it is not clear yet whether a market will survive under the present economic culture or not.

1.3 MARKET FOR BIO-BASED PLASTICS

Though small the bioplastic industry is a fast growing section of around 270 million tonnes plastic industry. The research activities from various markets estimate the growth of bioplastics ranging from 19% to over 30% a year. According to the opinion of Jim Lunt and colleagues the actual growth is nearly 19% value based as per a thorough evaluation of the actual volumes being sold as opposed to declared capacities.

Major parts of bioplastics comprise of the compounded starch products and first -eneration compostable bioplastics, for example, polylactic acid (PLA). They also have made a place in the market share from commodity thermoplastics, for example, polyethylene terephthalate (PET), polystyrene and polyvinyl chloride (PVC) in primarily single-use disposable application segments.

The first-generation compostable bioplastics have not penetrated the markets of polyolefin. This is because the polyolefin possesses the following properties:

- performance spectrum
- pricing compared to these bioplastics
- Extraordinary combination of low specific gravity

Recently, the non-compostable thermoplastics, for instance polyamides and thermoset products such as polyurethanes (PU), bio polyethylene (PE) partially bio-based polyesters such as PET and polybutylene terephthalate (PBT), epoxies, unsaturated polyesters (UP) and few other bioplastics are initiating their marketplace by directly replacing the oil-based equivalents cent percent.

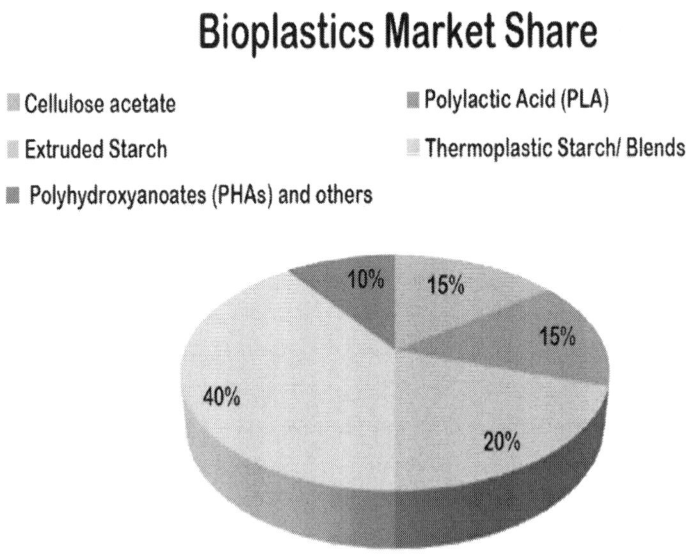

Figure 1.4: Market for bioplastic share.

Source: http://pubs.sciepub.com/jpbpc/2/4/5/

From the past five years, the market of bioplastics has seen a great rise. Earlier compostable bioplastics were used due to which the bioplastic market's performance suffered. But the additives and compounded polymer blends were used to overcome these performance deficiencies.

Though the opportunities to further enhance performance remains positive, keeping the cost and commercialization times for completely new approaches are still quite limited.

Subsequently, there still are significant issues that need to be solved for bioplastics against plastics that are petrochemical based. These issues include their pricing as well as their performance. Because of these issues, recently the commercialization activity in bioplastics on a large scale has shifted to the production of conventional monomers and existing oil-based plastics from renewable resource-based alternatives.

The bio-based polyethylene's breakthrough in the market in 2010 was successful in displacing their oil-based counterparts. Although, with the recent developments in the field of natural gas availability and efficiency has raised the questions over the ability of the bio-based olefin analogues to beat the products working on natural gas, in terms of pricing.

For example, monoethylene glycol, a bio derived chemical formula has successfully led to the commercialization of around twenty percent of renewable carbon content used in the manufacturing of PET by soft drink manufacturer giant Coca Cola in their "Plant Bottle".

And if we look at the future scope of bio-based products, Terephthalic acid, a bio aromatic component, possesses a huge potential as it has not been procured at a commercial level, but this is an area of immense development. As compared to the polyethylene segment, a large amount of natural gas can actually increase the rate of activity in the bio-aromatics field as aromatic compounds are not found in natural gas. In last decade, a lot of appreciative steps have been taken by the governments all over the globe in order to develop policy frameworks and strategies so that they can efficiently promote and support the Research & Development of competitive and sustainable bio-economy.

Many of these policies offer whole hearted support for the further improvements and development in the field of bioplastics and biomaterials, biochemical, expanding the scope of bio-based products or bio economy in general.

A lot of policies mostly focus on innovation and research. Many of the nations have already taken strict steps against plastics bags and implemented a complete ban on their usage. However, only a few countries have developed a specific set of policies targeting the development of bioplastics.

A lot of emphasis has been given on the replacement of food crops as feedstocks for bioplastics with forestry or agricultural biomass. A lot of Research & Development is taking place in technological advancements in order to extract sugar from these waste products in a cost-effective manner, but as of now, it's not available on a commercial scale.

The major issues faced with biomass feedstocks along with efficient extraction are the storage issues, logistics of supply, and accumulating enough quantities for making processes efficient and cost effective.

1.3.1 Conventional Plastics and their Market Size: Main target of Bio-based Compostable Plastics

Polystyrene

In developing nations like India, China, Saudi Arabia, Brazil and Iran, there is a huge demand for polystyrene and expandable polystyrene (EPS) and it is increasing rapidly with time, as per a research report by Global Business Intelligence Research.

The food service ware and polystyrene packaging industries seem to offer themselves as attractive and lucrative potential substitute markets for bio-based plastics like Polyhydroxyalkanoates (PHA's), polybutylene succinate (PBS) and Polylactic acid (PLA).

A market that is on direct target includes the food-service industry that requires a lot of single-use application products for example containers and hard trays, foamed cups, disposable eating utensils, bowls and plates. In the year 2015, it is estimated that global demand for disposables in the food service sector will grow by 5.4 percent per year and will reach around $53 billion.

As per the recent reports from RnRMarketResearch.com, the worldwide market size of biodegradable plastic packaging is estimated to reflect a cumulative average growth rate (CAGR) of 18.1% from 2013 to 2019. This will be corresponding to a value of around $8 billion. With the fact that consumers all over the globe are becoming more and more aware, along with contractual and general manufacturing activities in the region of North America and Europe, these developed geographies are estimated to project the maximum growth.

As per the reports, talking quantitatively, the U.S is the largest consumer of packaging products that are biodegradable in nature. The second position is held by Europe, with Switzerland, Sweden, Germany and the United Kingdom along with other nations in E.U playing a crucial role in exponential growth in the use of bio-degradable packaging products (Biodegradable Packaging Market: Global Industry Analysis and Opportunity Assessment ;2014 – 2020).

While food packaging takes the top position in the biodegradable packaging market, with more than 70% share by value of biodegradable plastic packaging and more than 40% share by value of biodegradable paper packaging, maximum growth in the future is expected to be from the beverage packaging application segment which is dominated by PET, PVC and polyethylenes.

The main issue with polystyrene is that there is no proper infrastructure to recycle it and that hurts the versatile nature of this material and is a major concern for its use, especially in the food industry. Polystyrene foam (EPS) has been banned in many geographies due to litter issues. A major reason for the fall of Polystyrene in food service ware industry is the fact that it can only be used once and it can contaminate the food, leading to possible discarding the products.

It has also been observed that Styrene monomer is a potential carcinogen for humans, irrespective of the fact that there is no proper proof or evidence present that supports this fact. However, this was sufficient to bring a change in consumer's perception that is motivating companies to find the possible replacements for food service ware products and polystyrene foam.

PVC

The Polyvinyl Chloride or PVC market was not well received since its inception and faced challenges for many years and 2014 has been no exception. Nevertheless, Suspension-grade PVC (S-PVC) has shown a rising scale and is expected to grow by 6 percent in 2014, reaching to 59 million tonnes per year as per the HIS Chemical. Companies were forced to trim the production to match the still-sluggish demand. Average of the operating rate of manufacturers around the globe is marked around 64.8 percent during 2014. This figure is a point lower as compared to that of in 2013.

It is expected that global PVC demand is expected to reach around 40.2 million tons in the year 2014. This figure is attained by a tenacious oversupply along with an annual growth rate of 4.15 percent. The PVC industry has remained fragmented in some regions like Western Europe, but there also, the signs of consolidation can be noticed.

One of the recent examples being the largest producers in the region, solVin (a joint venture between Solvay and BASF) and Ineos, has initiated talks to come together and merge their PVC business. As per the HIS, the S-PVC capacities of SolVin and Ineos are about 1.2 million tonnes/year and 1.8 million tonnes/year, respectively.

In other places around the globe, it is expected that PVC exports of North America will rise, the reason being the improved and stable cost position, while the mid-east will maintain its leading position as an exporter in 2014.

The main market segments that PVC enjoys include pipes and fittings, extruded profiles, and Flexible films. The major applications of PVC in these fields can be:

- Cable and Wire insulation
- Clear folders
- Water tubing, hoses, pipes
- Wallpapers
- Synthetic leather clothing mainly for use in diary covers and upholstery
- Door/window sidings and profiles
- Toys
- Laminates, extensively used in speaker boxes and gypsum boards
- Inflatable items like waterbeds or pool toys
- Clothing

Polyethylene Terephthalate (PET)

PET is seen as one of the major competitions for PVC and Polystyrene, although it does not fall directly in the category of commodity plastics. As per the latest research and market reports, it is expected that Polyethylene Terephthalate (PET) volume is anticipated to reach 69 million tonnes by 2016 due to increasing growth in fields like fibers, films and sheet extrusion markets. Containers, bottles and food packaging sectors currently account for about 62.6 percent of the PET market share. According to geographical analysis, it can be seen that the highest CAGR of 11.7 percent is expected to be the figures for Europe during the period of 2011 to 2016.

Europe is closely followed by the Asia Pacific region with the CAGR standing at 11.3 percent. It has been projected that America is expected to reflect a CAGR of 10.9 percent. The main reasons that made PET a success in these sectors are:

- Gloss and exceptional clarity
- Low water vapor diffusion, oxygen and carbon dioxide transmission compared to polyolefins. Recyclability.
- Ability to crystallize in orientation processes such as injection stretch blow molding and biaxial orientated films.

Starch based products like PHA's and cellulosic stand no competition against PET besides a minimal penetration by cellulosic into PET films.

As compared to PET, PLAs are not in a state to readily compete with it in the bottle market due to concerns about contamination of the PET recycle stream and poor barrier properties.

In many a place, PLA is replacing PET in fields like single-use thermoforming applications with a notable market size of approximately 7 percent of all PET packaging, and in numbers, it is just above one million tonnes worldwide, as per Smithers Pira, a UK based consultancy. PLA has also seen a widespread application in fibers mostly in wet wipes uses.

1.4 IMPACT OF BIO-BASED PLASTICS ON CURRENT RECYCLING OF PLASTICS

Bio-based plastics are already a part of consumption goods. Currently, the production of bio-based plastic includes ca. 1% of total plastics production and it is expected that there will be a rise in this share. It is estimated by the Nova Institute that this development of complete production of bio-based plastics will escalate by ca. 50% in 2021.

Therefore, the portion of bio-based plastics would then rise towards ca. 15%, the exact amount dependent on the progress of plastics based on fossils.

Crude oil was used as the feedstock of organic compounds. The search for alternative feedstock of organic compounds ends at the growth and development of bio-based plastics. It is important to look for an alternative to crude oil because not only it is a finite feedstock but also most products end as carbon dioxide in the air, resulting in global warming.

In comparison to this feedstock, bio-based compounds are quite sustainable; the source of prime raw material has a tendency to be renewed if enough care is taken during the growth of production and harvesting process.

There are a number of varieties of bio-based plastics. Further, they can be subdivided on the basis of their difference in the extent of biodegradability or on the basis of their similarity in molecules with prevailing fossil-based plastic. For instance, plastics such as bio-based PE (polyethylene) or bio-based PET (polyethylene terephthalate) are similar to their fossil-based counterparts PE or PET, because of this they are termed as 'drop-in' bio-based plastics.

Since the raw materials differ, there is a difference in the process of production of the building blocks of these plastics. Also, there are bio-based plastics which have building blocks of a certain basis that are more conveniently derived from feedstocks that are plant-based and also from which there is no development of fossil-based counterparts, as per their molecular chemistry.

Some examples of these types of plastics, which actually amount to becoming new products bearing new characteristics, are PEF (polyethylene furanoate), PLA (polylactic acid) and PHA (polyhydroxy alkanoate).

On the basis of their performance and not on the price alone, they give a chance to contest with fossil-based plastics. For instance, the use of PEF for the packaging of carbonated beverages. It is because of this it is expected that their market amount will bloom.

When the bio-based plastics will be extensively applied for some common uses, for instance, bottles, packaging, trays, etc. then they will also become a part of the waste stream and will enter the already known processes of recycling for fossil-based plastics. As discussed before, different types of bio-based plastics will be considered as new products. There may be chances of risks in some cases, from a definite slightest occurrence, they may come out to be mismatched with these processes, resulting in reduced quality of the recycled plastic stream in which the bio-based plastics have ended up.

In such a situation, it would affect the termination of material cycles in the process of recycling a plastic, which is particularly relevant given the current policy focus on the circular economy and on the recycling of plastics, as reflected in the recent launch of a strategy for plastics in the circular economy by the European Commission.

1.5 BIO BASED PLASTICS AND THEIR ENVIRONMENTAL IMPACT

1.5.1 Bio based Plastics versus Food

The relation between bio-based plastics and food can be seen as a dual relation where the bio-based plastic and food can compete for the similar feedstock and bio-based plastic can also be utilized in packaging of food.

The bio-based plastics versus food bout mimics the same as fuel versus food discussion, although at a minor scale. In around 2008, at the time when food prices shoot rapidly, as per Mitchel D, "the use of edible feedstock for the production of biofuels was strongly criticized for being a large driver for the food price rise".

As per OECD, 2008, a more subtle influence on food prices due to the production of biofuels has been reported. At the time of Dutch Food – Fuel debate during 2013- 2014, a mutual consensus among NGOs, businesses, government authorities and knowledge institutions was drawn such that "biofuels can offer opportunities as well as threats to food security, and that the ultimate effect strongly depends on how biofuels are produced" (2014).

According to a recent study by Kline at al (2016), it has been found that a healthy correlation between food and biofuels is possible only when the correct measures are taken into consideration, and the biofuel produced may act as a moderator to bring stability in food prices by delivering farmers a more trustworthy and secure market place and further leading towards biofuel production that is more sustainable.

As per IfBB, 2015, "The present worldwide production capacity of bio-based and biodegradable plastics lies around 1.7 Mtonne in 2014" and according to Piotrowski et al, 2015, "Compared to the total amount of biomass produced and harvested each year, which lies around 11.5 Gtonne", which is very small.

Further as per European Bioplastics, 2015, they have calculated that the land needed to grow the feedstock for bio-based plastics amounted to 0.01% or the agricultural area in 2013 that may grow to 0.02% by 2018. The growing demand for feedstock is directly proportional to the growing demand for bio-based plastics.

As per the research by Bos and Sanders (2013), it can be evaluated that even if we would vile all current fossil plastics production around the globe on biomass instead of feedstock, the total rise that is expected in the demand of feedstock will be in the vicinity of 5 percent of total quantity of biomass harvested and produced every year.

But this is very less likely to happen as it is estimated that the manufacturers are extensively working on developing new and better alternative feedstock such as lignocellulosic feedstock used as chemical building blocks, currently being examined further by manufacturing firms like Biochemtex, settled in Italy and Avantium, to name a few among many others.

A lot of R & D is going in the field to directly utilize the methane and carbon dioxide in fermentation that can play an important role in lifting off the pressure on feedstock. As per the Elbasan and Bos, 2016, "the recent insights on the possibility of sustainable co-production of biofuels and food will also applying this case, providing an additional market for biomass, next to food, feed and energy".

As per the recent reports from Bio Based Press, 2016, a large amount of bio-based plastics are currently developed by using the starch or sugar, a lot of research and development has gone into this by both the knowledge institutes and industry in order to implement desired molecules from the side stream of the agro industry for example, beet pulp can be used efficiently for the production of bio-based plastics. Cosun among others is very active in this field

Particularly, bio-polyesters are looked upon as an interesting option for the manufacturing of these types of feedstock. This will directly provide the high valued output to the agro industry of their side-streams and also enable the manufacturers to reduce their dependency on food-feedstock demand for developing bio-based plastics.

Biodegradable and Bio-based plastics together form an exciting packaging material for foods. For example, in case of food and beverages in in-flight meals, that includes cutlery and plates, biodegradable plastics can be composted together with the meal remains, which saves the necessity to clean and dry. And this in return, leads to a favorable LCA.

In further details, it has also been found that some biodegradable and bio-based plastics, like PLA, possess barrier properties which play an important role in keeping food stay fresh for a long time. For example, the shelf life of greens like lettuce can be extended up to two days by simply packing it in PLA.

It is a well-known fact that approximately 30 percent of the total food production gets wasted, and mostly due to the passing of the sell by date which either happens in retail markets or a part of that also happens at home. With a mere extension of the shelf life of products that are perishable, by the application of biodegradable and bio-based plastics, rising problem of food wastage around the globe can be minimized. According to a rough evaluation by Bos, 2016, it has been found that lettuce wastage can be reduced by up to 10 percent just by changing their packaging, which in turn helps in saving land use due to a reduction in wastage of food.

This reduction of food waste outweighs the additional land use for the production of the packaging. This is applicable in the cases of the first generation of feedstock, and its application for the side stream for the plastic production will further help in even additional favorable results.

1.5.2 Environmental impact Comparison between Bio-based Plastics and Fossil-based Plastics

The manufacturing of fossil plastics and production of bio-based plastics have altered reaction on various impact categories. Greenhouse gas emissions and non-renewable energy are commonly more desirable for bio-based plastics and this replacement of fossil plastics by bio-based plastics thus leads to lower non-renewable energy application and GHG emissions.

The usage of land for the development and manufacturing of feedstock for bioplastics may, in some cases, lead to indirect or direct change in the pattern of land use, which can have a negative impact on the GHG emission reduction.

"The GHG emission reduction that is reached by bioplastics is generally significantly larger than that by biofuels, the relative impact of ILUC on GHG emissions will ,therefore, be smaller for bioplastics than for biofuels" (Bos et al, 2010; Bos et al, 2012).

Considering the different aspect of environmental impact, for the sectors related to agriculture, such as acidification and eutrophication, bio-based plastics commonly leave a deeper impact as compared to that of fossil plastics. As there is a considerable difference in the effect that both bio -ased plastics and fossil plastics have, no particular rule can be given in this direction.

1.5.3 Biodegradable and Bio-based Plastics from Genetically Modified Organisms

Genetic Engineering (GE), also called Genetic modification, can be used to modify the size and composition of plants, enhance their fruit yield and improve their resistance against insects and pesticides. (Gould, 2016).

Although, genetic engineering has been facing a heated resistance from social groups working for environmental protection that are raising their voice against the concerns and potential risks that are related to the application and usage of genetically modified organisms (GMO) like unintentional spreading of GMO to conventional crops, human health effects, insect resistance and diversity.

A US group of researchers and scientists, with researchers taking part from Mexico and Netherlands, has recently published a research paper stating that they have "found no cause-and-effect relationship between GMO crops and environmental effects. However, the complex nature of assessing long-term environmental changes often made it difficult to reach definitive conclusions" (Gould, 2016). Further, in a research review by European GMO researchers, they have stated that "25 years of field trials have not shown evidence of environmental harm" (Van Montagu, 2010).

The other side of the story includes Greenpeace that discussed examples of GM crop research and the study highlighted the negative effects that GMO have on non-target species (insects) and beneficial insects, while taking into account the results of the research that indicates cumulative and toxic impact on aquatic and soil ecosystems, and also states that new insects are filling the blank space left due to the absence of rivals, that were earlier taken care by GM crops (Greenpeace, 2011).

To summarize, it may be said that there still exist some societal and scientific clashes about the impact of GMO especially in case of open ground cultivation methodology. The main reasons that amplify the situation are:

- the potential risks are complex and possibly potentially huge as well
- it takes a long time before definitive conclusions on the use of a particular GMO crop can be drawn
- the potential gain is huge

It can be said that; GMO crops have no harm if they are cultivated after a proper risk analysis. As per ISAAA and FAO, the first GMO crop on a commercial level was cultivated in 1996; the area in which it was cultivated has steadily converted into a total GMO crop cultivated area. This area sums up to around 180 million ha in 2015, which is about 13 percent of the total arable land.

Further, in Europe, a very small and limited area has been exposed to GMO crops. According to Molenveld and Van den Oever, 2015, "PLA is produced based both on GMO and non-GMO corn and non-GMO sugarcane". It is not required for GMO crops to produce PLA. For the production of bio-based plastics, it is not required to use genetically modified (micro) organisms (GMOs). For example, many monomers like 1, 3-Propanediol and succinic along with a polymer like PHA can be developed by using fermentation mention using GMO strains. Since fermentation generally takes place in a closed environment, the GM organisms can be observed and controlled (white biotechnology). Also, in a closed system, GMO strains are extensively used in performing the conversion of woody biomass to fermentable sugars for the production of second-generation polymers.

1.5.4 Disposing off in the Environment

- The problem of litter is vast and growing, hence, biodegradable plastics are not and should not be taken as a solution to this problem. Littering is highly criticized in all forms of waste, be it in the sea or on the land, including all kinds of plastics.
- As an alternative, this matter requires to be taken care of by informative and educative methods to create and spread awareness to properly curb and control the disposal, recycling and management of waste.
- Before designing any product, proper and efficient recovery measures must be thought of. Considering the situation of biodegradable plastic items, organic recovery is a viable alternative besides energy recovery options and regular material.
- It should be taken into account that designing a product "for littering of any kind" can be misleading and encourage the misuse of disposal, which will defy the mere origin of bio-based plastics. Concluding the remark, biodegradability does not establish a license to litter.

1.6 FUTURE TRENDS OF BIO-BASED PLASTICS

Plastics that are obtained from the biomass sources which are renewable are termed as bioplastics. Few examples of renewable biomass sources are microbiota or corn starch, vegetable oils and fats, etc. Biomass is basically an agricultural byproduct which is used to make bioplastic. Bioplastic is misleading because it suggests that any polymer derived from the biomass is environmentally friendly. The application of the expression "bioplastic" is rejected.

It is advisable to use the expression bio-based polymer in place of bio-plastics. More amount of fossil fuel is needed for the production of conventional plastics (origin is petroleum). They also add more greenhouse gases to the atmosphere as compared to the production of bio-based polymers (bioplastics).

Not all bio-based are degradable. Environmental and economic concerns have driven the development of bio-based polymers and materials in recent years.

Most of the efforts made are concerning the production of biopolymers directly and the manufacturing of bio-based monomers for their polymerization by biotechnological or chemical routes. In various forums, bio-based products are pushed by Mark Zuckerberg, Bill Gates and Obama.

- The 44th U.S President Barack Obama announced on October 27, 2014, that BIO-BASED Materials are one of three emerging technologies for U.S. competitiveness.
- In November 2015, at the Paris Summit, Mark Zuckerberg and former President Obama along with 150 leaders of the world agreed to control the climatic conditions by reducing the applications of fossil fuels.
- In September 2016, Bill Gates capitalized in Plant rose technology which will facilitate cost-advantaged building blocks for polymers and chemicals.
- Mainly most of the corporations around the world are supporting the progress in Environmental Impact and Sustainability in several ways, but some of the corporation's break-through concepts have begun to materialize lately.

While purchasing a bio-based plastic, one should be certain about two things. One that it should be certified and confirmed that the product is produced using biomass and second, how much bio-based plastic is present in the product. Petroleum based products and products derived from biomass are identical as they have similar chemical and physical properties if their molecular structure is the same.

Thus, in order to increase general consumer knowledge and encourage bio-based plastics, the amount of synthetic polymer present in a product shall be a plastic product of 25.0 wt. % or more in one of the authentication conditions of the aforementioned system.

The degree of bio-based synthetic polymer is the ratio of the biomass origin resin to the total plastic product. The Japan Bioplastics Association (JBPA) is handling the Biomasspla, under the same system, a product that fulfills the required standards are certified as Biomasspla and permitted to use Biomasspla.

Plastics are the products that are manufactured using raw materials like pigments, additives, polymers and fillers. The source of these raw materials can be either biomass or petroleum. Therefore, it is mandatory to find out the exact amount of plastic content from a biomass source in the provided plastic. A great amount of work has been done to estimate bio-based carbon ratios for various polymeric composites with additives and fillers and discussed the repeatability and accuracy of this evaluation method. They have used accelerator mass spectrometry (AMS). In order to find out the amount of biobased plastics in provided plastics, the ratios of carbon in plastics can be predicted by the ratio of ^{14}C to ^{12}C that is measured using Accelerator Mass Spectrometry (AMS) conforming to the standard ASTM D 6866 "Standard Test Methods for Determining the Bio-based Content of Natural Range Materials Using Radiocarbon and Isotope Ratio Mass Spectrometry Analysis."

The principle of using ^{14}C in this method is derived from the method of dating measurement for various historical materials in archeology (Jull and Burr 2006; Currie 2004). The radioisotope of carbon atom is ^{14}C which has a half-life of 5730 years. ^{14}N atoms continuously generate ^{14}C carbon atoms because of their interactions with cosmic radiations in the atmosphere.

The ratio of ^{14}C to ^{12}C in modern air is constant at approximately 1×10^{-12} in spite of the period. Plants use carbon dioxide present in the atmosphere, as a raw material for preparing food by the process of photosynthesis. In plants, the ratio of ^{14}C to ^{12}C is 1 x 10^{-12} just after the process of photosynthesis.

In plants, the ^{14}C slowly decays and gets converted to ^{14}N. Since the number of ^{14}C atoms reduces continuously and becomes half after 5730 years, the lifetime of materials comprising carbon atoms can be judged by the ratio of the number of ^{14}C atoms to the number of ^{12}C atoms and the half-life of ^{14}C.

The ratio of ^{14}C to ^{12}C is measured using AMS. This ratio is quite low, approximately 1 x 10-12. In accordance with formulas for radiometric dating in ASTM D 68866-08, 1950 was defined as the standard year. In ASTM D 6866-08, formulas for radiometric dating are used for determining of the bio-based carbon content.

By comparing the measured ratio of ^{14}C to ^{12}C, a percent modern carbon (pMC) value can be estimated and the standard ratio of ^{14}C to ^{12}C can be found by the suitable primary reference (oxalic acid) of SRM 4990c supplied by National Institute of Standards and Technology (NIST), USA (SRM 2013). Theoretically, biobased carbon ratios for petroleum-based materials are estimated at 0%, and for bio-based materials at 100%.

As predicted by analysts the global market of bioplastics will rise at a CAGR of 29.3% during the years 2016-2030. One of the reports reveals, the rise of raw materials that are renewable and bio-based will come out as the main growth progressing feature for this global market. The global market of bioplastics is developing by the rise of raw materials like vegetable crop derivatives and starch, which are biomass, renewable resources and bio-based.

In the year 2015, the bio-based plastics contributed more than 80% of the global market of bioplastics. Bioplastics have a number of applications such as domestic goods and packaging, because of which the manufacturers of plastics have reduced the overdependence on petroleum-based plastics.

At present, the trend that is influencing the bio-based plastic market is that the bio-based raw material is available at a low cost. These cheap raw materials are either the ones produced from non-food products or waste leftovers of food products which serve as a great alternative feedstock for the production of bioplastics.

The use of non-food raw materials like biomass and organic waste for the manufacture of bioplastics also helps in waste management. The widely used bio-based plastics are polymers like Bio polyamide, Starch, Polylacticacid (PLA), Bio PET, blends of PLA, Bio Polyethylene, blends of starch, Bio Poly propylene etc.

By 2020. It is estimated that the food service and packaging industry will cover approximately 69 percent of the total market space in order to be a dominant stakeholder in the global market of bioplastics. The major usage of bioplastics can be seen in the production of loose-fill, bottles, bowls, cups, pots, flexible films and many more items.

Rigid and flexible packaging are the two most commonly used types of bioplastics packaging. Packaging items that are made of bioplastics in utilized in the packaging of dry snacks, bakery goods, fresh foods, candy, meat trays, juice bottles along with the coatings for beverage cups, card stock and film coatings.

1.6.1 A Central Role for Bio-based Plastics in a Future Bio Economy?

Bio-based and biodegradable plastics are replacements for petroleum-based plastics, as a solution in the struggle against climate change. However, estimates of GHG emissions savings from the production of various bioplastics and bio-based chemicals vary widely; an unhelpful situation for the industry. In this case, a target for policy action, certainly regarding the methods for making calculations, which are done by Life Cycle Analysis (LCA). The earlier bio-based plastics were biodegradable, but the improved bio-based plastics are not only biodegradable but also comparatively more durable. Thus, there is a noticeable shift in the market towards the improved bio-based plastics. But nonbiodegradable substitutes for the fossil-derived thermoplastics are produced in massive volumes, especially polyethylene (PE), polypropylene (PP), and polyethylene terephthalate (PET).

Recently, it was predicted that the global production volume for bio-based plastics will escalate from round about 1.2 million tonnes in 2011 to approximately 5.8 million tonnes by 2016. So far, the most powerful development has been in the bail-based, non-biodegradable bioplastics discussed above. Perfecting these in bio-based forms dramatically increases the uses of, and markets for, bio-based plastics.

Theoretically, their GHG emissions should be significantly lower than their Petro-equivalents, a theory that needs to be tested on a case-by-case basis. Little or all of the carbon in the bio-equivalents comes from the atmospheric carbon dioxide fixed during the growth of the plants used as biomass, and thus they are near to the carbon neutrality in comparison to the Petro-equivalents when their carbon comes back to the atmosphere as carbon dioxide.

Since the molecules of bio-PP, bio-PE, bio-PET and upcoming bio-PVC are similar to the molecules of the Petro-equivalents, therefore their performance properties must be similar. This also makes way for bio-based thermoplastics.

There is no impediment for them to enter the existing plastics recycling infrastructure, whereas other bioplastics such as polylactic acid cannot readily do this. In some countries, incineration with energy recovery is a striking end-of-life option with the included environmental benefit of electricity production from waste. When cradle-to-grave LCA is performed, the end-of-life efficiency is important to allocate the environmental performance of a plastic molecule.

Composting of biodegradable plastics at industrial-scale is possible only in some countries. Even in the developed countries of Europe, the biodegradable and bio-based plastics industry faces the challenge as composting along with the use of compost as fertilizers is not allowed, even when the products comply with the strict criteria of the EN 13432 composting standard.

The mixture of policy in this space is complex and requires to be revised, else the benefits of biodegradable plastics will be worthless. The alternative of anaerobic landfill disposal nullifies biodegradation in most cases and does not address the landfill dilemma.

1.6.2 A Vital Role of Bio-based Plastics in the Bio Refineries of the Future?

A lot of oil refineries that produce diesel and petrol delivers on extremely low profits. They gain and increase profit margins by bringing together fuel and chemical production within a single operation. In petrochemical oil refineries, approximately 7 to 8 percent of crude oil is dedicated to chemical production that directly results in an annual profit of around 25 to 35 percent.

As compared to conventional oil refineries, bio-refineries similarly face the same market dynamics, as in most of the locations the production cost of biofuels in biorefineries is significantly higher than that of diesel and petrol.

With the plastics in this market depicting a central production volume between low-volume chemicals and high-volume fuels, the bio-based plastics are seen as a center to the economics of the integrated biorefinery. Furthermore, the largest production volume bio-based chemicals can be used as monomers to produce the plastics.

REFERENCES

1. Alaerts, L., Augustinus, M. and Van Acker, K. (2018). Impact of Bio-Based Plastics on Current Recycling of Plastics. *Sustainability*, [online] 10(5), p.1487. Available at: https://www.mdpi.com/2071-1050/10/5/1487/pdf [Accessed 28 Mar. 2019].
2. Chemarc.com. (2017). *Bio Based Plastics- Future trend.* [online] Available at: https://www.chemarc.com/content/article/bio-based-plastics-future-trend/58a571169855f3473492e406 [Accessed 28 Mar. 2019].
3. der Zee, M., Bos, H., den Oever, M. and Molenveld,, K. (n.d.). *Bio-based and biodegradable plastics – Facts and Figures Focus on food packaging in the Netherlands.* [ebook] Available at: https://www.wur.nl/upload_mm/1/e/7/01452551-06c5-4dc3-b278-173da53356bb_170421%20Report%20Bio-based%20Plastic%20Facts.pdf [Accessed 28 Mar. 2019].
4. Green Dot Bioplastics. (2019). *Bioplastics 101: An introduction to key terms in sustainable plastics - Green Dot Bioplastics.* [online] Available at: https://www.greendotbioplastics.com/bioplastics-101-introduction-key-terms-sustainable-plastics/ [Accessed 28 Mar. 2019].
5. Marketplace Opportunities for Integration of Biobased and Conventional Plastics. (2014). [ebook] Available at: https://www.auri.org/assets/2014/09/AIC185.biobased1.pdf [Accessed 28 Mar. 2019].
6. Omicsonline.org. (2016). *Methods of production of Bioplastics |.* [online] Available at: https://www.omicsonline.org/conferences-list/methods-of-production-of-bioplastics [Accessed 28 Mar. 2019].
7. Philp, J., Ritchie, R. and Guy, K. (2013). Biobased plastics in a bioeconomy. *Trends in Biotechnology*, [online] 31(2), pp.65-67. Available at: https://www.researchgate.net/publication/234699869_Biobased_plastics_in_a_bioeconomy [Accessed 28 Mar. 2019].
8. Philp, J., Ritchie, R. and Guy, K. (2013). Biobased plastics in a bioeconomy. *Trends in Biotechnology*, [online] 31(2), pp.65-67. Available at: https://www.cell.com/trends/biotechnology/fulltext/S0167-7799(12)00204-1 [Accessed 28 Mar. 2019].
9. Storz, H. (2014). *Bio-based plastics: status, challenges and trends.* [online] Available at: https://www.researchgate.net/publication/265433282_Bio-based_plastics_status_challenges_and_trends [Accessed 28 Mar. 2019].

10. Sutherland,, I. (n.d.). *BIOPLASTIC AND BIOPOLYMER PRODUCTION*. [ebook] Available at: http://www.eolss.net/sample-chapters/c17/E6-58-05-05.pdf [Accessed 28 Mar. 2019].

Chapter 2

STARCH FILLED PLASTICS AND MODIFICATIONS

CONTENTS

Starch is a type of natural biopolymer which has two types of glucose polymers named as amylose and amylopectin. Starch is beneficial for producing plastics because of its abundance, renewability and biodegradability. The ever-lasting pursuit of producing starch-based plastics has observed different kinds of starches such as granular starch, plasticized starch, modified starch, both biodegradable and non-biodegradable. The chapter initially deals with the brief introduction of starch-based plastics and preparation of biodegradable plastics from tuber and root starches. The structure and properties of starch-based plastics has also been described. At the end of this chapter, applications and influence of the starch-based plastics are described along with the future aspects of these plastics in the field of packaging.

2.1 WHAT ARE STARCH BASED PLASTICS?

Big manufacturing companies need to stay in the competition for their survival. For this, they need bioplastics. Bioplastic is made up of chemical compounds that are derived from or synthesized by microbes such as bacteria or by genetically modified plants.

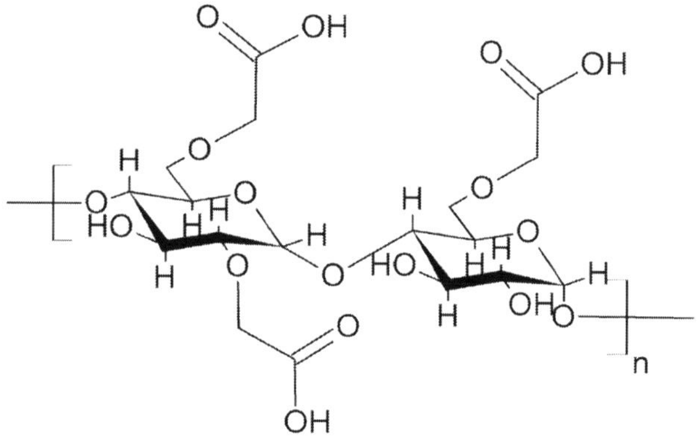

Figure 2.1: Structure of modified starch.

Source: https://en.wikipedia.org/wiki/Modified_starch

Starch-based plastics can be easily incorporated with various petroleum-based polymers or biopolymers and create unique composite materials. That is why they can be used in a variety of applications. These composite materials can then be injection molded or extruded using standard processing machinery.

In comparison to alternative bioplastics, starch-based plastics are cheaper. They have a wide range of physical properties that alternative bioplastics lack, such as tensile strength and heat tolerance. Starch composites can also incorporate recycled plastics.

Since starch-based plastics can replace petroleum-based polymers with natural ones, they can be used to reduce the carbon footprint of traditional resins. It is also highly degradable, meaning it can be used alongside a compostable polymer without interfering with the degradation process.

Numerous efforts have been made to use starch to make plastics but none of them has borne fruit. They have failed to make a mark in the market mainly because better and lower cost plastics can be made from other raw materials, particularly petroleum.

Although the demand for plastics is increasing day by day, the industry is forced to consider starch in plastics that are biodegradable and flame resistant mainly because of the decreasing availability and increasing prices of conventional raw materials.

The rigid urethane foam market was using around 100 million pounds annually of starch-derived products, such as sorbitol, methyl glucoside, and glycol glycosides. Currently, only 1 to 2 million pounds of sorbitol is entering this market. In the next few years, it is predicted that major growth will be seen in the consumption of disposable packaging, food trays, plates, and eating utensils.

Plastic films are widely used in agriculture but since they are nonbiodegradable, it causes problems. These films are needed to provide weed control, conserve moisture and nutrients, and warm the soil for early crop production. To overcome this problem, starch material is incorporated in polyurethanes as filler of resins and to improve flame resistance of plastics.

2.2 STARCH PLASTICS (THERMOPLASTIC STARCH)

Sustainable plastics can be developed through starch and its blends with aliphatic bio polyesters and cellulose derivatives. Starch has many advantages such as

- It is completely biodegradable
- It is available abundantly.
- It is inexpensive

- It is regenerated from carbon dioxide and water by photosynthesis in plants.
- However, unmodified starch-based plastics have poor physical properties.
- Following are the disadvantages of unmodified starch-based plastics:
- They are hydrophilic and dissolve easily in water
- They have poor mechanical properties when moist
- They are too brittle when they are dry
- They have a strong tendency to recrystallize and shrink noticeably when drying.

Starch-based films, composites, and glues have a wide variety of applications including automotive, construction, packaging, marine, electronic and aerospace industries, therefore, various studies have been conducted to prepare them with improved properties.

Being brittle is the main disadvantage of starch. Its brittleness can be decreased by blending with various natural plasticizers such as glycerol, glycol and sorbitol and by ester- or etherification. Unfortunately, these blends and modifications have poor dimensional and thermal stability and low mechanical strength However, this can be overcome by grafting multifunctional monomers onto the polymer backbone and by subsequent crosslinking.

The main agents used for rafting and crosslinking are phosphoryl chloride, acid anhydrides, methacrylate's, epoxies, epichlorohydrin, glyoxal, and acrylonitrile among many other compounds. These chemical modifications help in making starch insoluble in water and improving its stiffness and tensile strength. But this results in making them harmful for the environment.

An environmentally friendly crosslinking reaction is the esterification of starch with naturally occurring or bio-based acids such as citric, succinic or itaconic acid which react with multiple hydroxyl groups at elevated temperature, thus esterification takes place during the drying stage of the mixture (film).

The blends usually contain glycerol or other polyols which also react with the diacids, i.e. polyol acts as both a chain extender and plasticizer. As we have already mentioned that they have the disadvantages of low resilience, high moisture sensitivity and high shrinkage, these disadvantages can be overcome by blending them with natural and synthetic polyesters such as polylactic acid, polycaprolactone, and polyhydroxy butyrate.

Compatibilizers such as PVA and starch-g-polymers are often added to improve the compatibility of the starch/polyester which also improves the mechanical properties. These approaches do not affect the biodegradability of starch and many of the compositions are fully compostable. They also have improved impact resistance and dimensional stability. However, polyester-starch blends are less strong than cross-linked starch.

To further improve the biodegradability of commodity plastics such as polyethylene, polypropylene and polystyrene, granular starch has also been used as a filler to improve the compatibility with polyolefins, the starch granules are usually surface treated or chemically modified to produce hydrophobic starch.

Starch can be easily used with any strong hydrogen bonding compound such as poly (ethylene-co-vinyl alcohol) and/or poly (vinyl alcohol). These compounds can also function as compatibilizers for polyester-starch blends. Typical blends consist of starch, PVA (or copolymer), glycerol and urea. These compositions have the advantage of being fully biodegradable and have mechanical properties between those of LDPE and HDPE.

Another approach uses copolymers of olefins and polar monomers such as (meth)acrylic acid, the later acts as a compatibilizer. Thermoplastic blends of up to 50% starch and poly (ethylene-co-acrylic acid) (EAA) have been prepared. These difunctional reagents reinforce granules by reacting with more than one hydroxyl group, which helps in crosslinking the starch.

To improve the properties and lower the cost of modified and unmodified starch, it is often blended with other bio-based polymers. Films made from these plastics are often transparent, flexible, and have good or acceptable physical properties.

2.3 BIO-PLASTICS FROM STARCH BASED FOOD CROPS

It is believed that bioplastics or bioorganic plastics will play an important role in the packaging sector in the near future. This is because they are derived from renewable biomass sources such as vegetable oil, corn starch or microbiota in contrast to fossil fuel plastics which are derived from petroleum residues. Bio-plastic polymers have great potential to contribute to material recovery, reduction of the landfill, customized products and use of renewable resources. Bio-plastics represent a new material group, which can make use of all the established recovery and recycling technologies for conventional plastics and moreover offer the new option of recycling bioplastics have the following advantages over petroleum-based plastics:

- They consume fewer fossil fuel resources because no fossil fuel feedstocks are used.
- They emit less carbon dioxide over their life cycle.
- They consume less energy to produce.
- They have fewer health concerns associated with them.
- They are generally compostable.

All these traits would encourage the transition to bio-plastics. Bioplastics are still in their infancy with low market volume, so we still have time and opportunity to develop solutions. Recycling issues should not lead to hampering the development of bioplastics.

Landfilling untreated organic waste needs to be discontinued and onsumers should be encouraged to collect their organic wastes separately in compostable bags. This would help solve the problem of methane emissions and protect ecology and environment.

Bioplastics are going to be very useful in the future in the following ways:

- They can be used in food packaging.
- They may also form major components in electronic housings and vehicles.
- They can be used in medicine delivery systems and chemical microencapsulation.
- They may also replace Petrochemical based adhesives and polymer coatings.

However, the plastic market has its own difficulties. It is complex, highly refined and manufacturers are very selective about the specific functionality and cost of plastic resins. Bioplastics will need to be more cost-competitive and provide functional properties that manufacturers require in order to make their own position in the market. There are several surveys concerning starch for the production of biodegradable plastics. Nevertheless, starch limitations with reference to its moisture resistance and its mechanical properties restrict its use significantly.

To overcome these problems, starch is used in mixtures with plasticizing elements and other biodegradable polymers. However, these mixtures raise the cost of the products because biodegradable polymers are more expensive, therefore reducing the consumption in some market niches. Cost reduction implies additional research and increase in production. In the last few years, an alternative has been found. Biopolymers derived from fermentation such as polylactic acid (PLA) produced from sugar or starch, have been in an outstanding position. Since they have better resistance to moisture, they are suitable for use in several markets.

polylacticacidBesides PLA, many companies have made bulk investments in bio-based technologies, with the aim of getting polymers directly from the plants. It will allow cost reductions and significant energy savings, making the production of polymers feasible.

2.4 STARCH-POLYMER COMPOSITE

Starch is an abundant, inexpensive, natural raw material which permits the development of products recyclable to atmospheric CO_2 when biodegraded or incinerated.

This chapter reviews the main results obtained in the fields of starch-filled plastics and thermoplastic starch, paying particular attention to the concepts of gelatinization, de-structurization, extrusion cooking, and the use of complexed starch in specific synthetic polymers.

Aspects such as processability, the physicochemical and physio-mechanical properties, and the biodegradation behavior of starch-based materials on the market are briefly considered.

2.5 PREPARATION OF BIODEGRADABLE PLASTIC FILMS FROM TUBER AND ROOT STARCHES

At present, we know that usage of synthetic plastic can lead to some serious environmental problems because they are non-biodegradable. People are continuously being made aware of the environmental pollution and as a result, there has been a paradigm shift in the use of biodegradable materials especially from renewable agriculture feedstock and food processing industry.

Most of the materials used for packaging purposes have been developed over a period of 50-60 years and are durable and inert in the presence of microorganisms which lead them to survive for long terms. However, keeping in mind the current situation of environmental pollution, it is very necessary that we shift to biologically degradable polymers

Governments in developing countries have been supporting a lot in this. Several studies have been performed to analyze the properties of starch based biodegradable plastic. Starches from roots and tubers (Irish potato, cocoyam, cassava) were studied in their slurry, powdered, gelatinous and crystalline form.

Amylose is responsible for film forming capacity of starches and starch contains around 30% of amylose making it a good raw material to prepare biodegradable film plastics. It has the following advantages:

- It is a renewable source.
- It is widely available.
- It is easy to handle.
- It is inexpensive.

Biodegradable plastics are plastics that will decompose in natural aerobic (composting) and anaerobic (landfill) environments. Biodegradation of plastics occurs when microorganisms metabolize the plastics to either assailable compounds or to humus-like materials that are less harmful to the environment.

They may be composed of either bioplastic, which are plastics whose components are derived from renewable raw materials or petroleum-based plastics which contain additives. Polymer additives play a vital role in modern plastics.

- They perform the following role:
- They overcome obstacles in processing.
- They increase material durability.
- They help product designers obtain the trendy looks feel and performances that their consumers demand.

As per the set local regulations, additives can also be used. Inorganic or inorganic compounds are blended in plastic. The number of additives ranges from 0% for polymers used to wrap foods to more than 50% for certain electronic applications. The average content of additives is 20% by weight of the polymer.

It was in the 1970s that starch was developed as a cost-effective additive to synthetic plastic. At that time, it was realized that standard starch was not suitable. The moisture content of the starch needed to be reduced so that it could be processed in polymer melts above 160C.

The starch/polymer interface can be modified by making the normally hydrophilic starch surface hydrophobic. This proved to be beneficial. However, photooxidation of polymers occurs which may lead to various effects such as oxidation, chain scission and uncontrolled recombination and cross-linking reactions. Plasticizers are additives that increase the plasticity or fluidity of a material.

The following are the advantages of using biodegradable plastics:

- They decompose to environmentally friendly composites.
- Most biodegradable plastics contain no allergens and are safe for atopic consumers.

They are non-toxic. They are made from natural elements, so they contain no chemical to exude toxic and poisonous wastes while breaking down in compost.

It has the greatest advantage of having the potential to rely less on oils which is synonymous to energy conservation.

A significant amount of oil used to produce plastics will be conserved. Our dependency on oil sources can be reduced and a considerable amount of energy can be saved if we produce biodegradable products made from local biomass materials.

In the long run, adapting to using biodegradables can lead to domestic solutions. Biodegradable plastics are easier to recycle because they are made from materials that are fully biodegradable. Biodegradable plastics can be used and reused more efficiently even as household utensils and in restaurants.

The only major drawback of biodegradables is that if they are not disposed of properly, it can lead to an inefficient breakdown which can release toxins into the environment. These toxins may include methane and carbon dioxide, both of which contribute to the greenhouse effect.

The most efficient way to dispose of biodegradable plastics is by composting. Biodegradable plastics are made from organic sources which include Corn and soybeans. Currently, research is being carried out to compare the proximate analysis and functional properties of the native starches isolated from three different tubers which are Irish potato, cocoyam and cassava.

2.6 STRUCTURE AND PROPERTIES OF STARCH-BASED PLASTICS

Starch is mainly composed of two homopolymers of D-glucose: amylase, a mostly linear αD (1, 4')-glucan and branched amylopectin, having the same backbone structure as amylose but with many α-1, 6'-linked branch points.

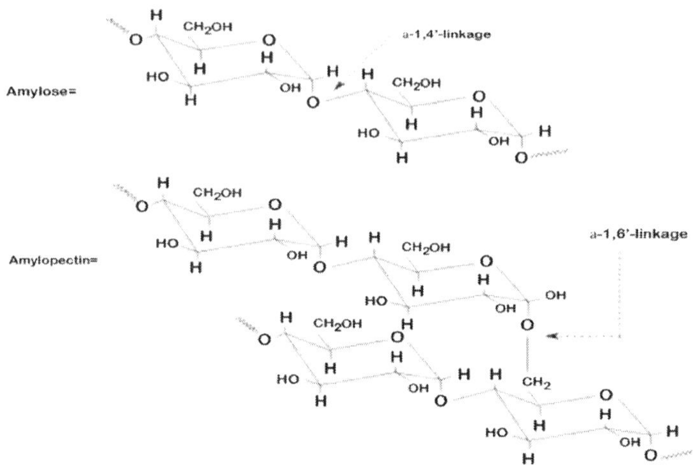

Figure 2.2: Molecular Structure of Starch.

Source: http://www.expresspolymlett.com/letolt.php?file=EPL-0000946&mi=dc

There are a lot of hydroxyl groups on starch chains, two secondary hydroxyl groups at C-2 and C-3 of each glucose residue, as well as one primary hydroxyl group at C-6 when it is not linked. Evidently, starch is hydrophilic. The available hydroxyl groups on the starch chains potentially exhibit reactivity specific for alcohols.

- They have the following features:
- They can be oxidized and reduced.
- They may participate in the formation of hydrogen bonds, ethers and esters.

Amylose and amylopectin can be found in starch in different proportions ranging from about 10-20% amylose and 80-90% amylopectin depending on the source.

Amylose is soluble in water and forms a helical structure.

The shortbranched amylopectin chains are able to form helical structures which crystallize, as a result of which starch occurs naturally as discrete granules. Starch granules exhibit hydrophilic properties and strong inter-molecular association via hydrogen bonding formed by the hydroxyl groups on the granule surface.

Owing to its hydrophilicity, the internal interaction and morphology of starch will be readily changed by water molecules, and thereby its glass transition temperature (Tg), the dimension and mechanical properties depend on the water content.

Tg of native starch can be as low as 60 to 80°C when the weight fraction of water is in the range 0.12 to 0.14, which allows starch to be successfully injection molded to obtain thermoplastic starch polymers in the presence of water.

There is a constant need to improve the degradation rate of some degradable hydrophobic polymers, which can be done by the hydrophilicity of starch. Starch is totally biodegradable in a wide variety of environments.

It can be hydrolyzed into glucose by microorganism or enzymes, and then metabolized into carbon dioxide and water. It is worth noting that carbon dioxide will recycle into starch again by plants and sunshine. Native starch is not used directly because of its following drawbacks:

- It is poor in processability.
- It is poor in dimensional stability.
- It has poor mechanical properties.

2.7 MODIFIED STARCH & STARCH-BASED PLASTICS

Starch is one of the most abundantly found biopolymers. It is a known fact now that it is completely biodegradable, inexpensive, renewable and can be easily chemically modified. Therefore, it is not surprising that starch and its derivatives have received increased attention as biodegradable alternatives to conventional petroleum-based plastics. Maize, potato, tapioca and wheat starch are the most abundant and cheapest starches.

Like cellulose, starch can be considered a condensation polymer because its hydrolysis yields glucose molecules:

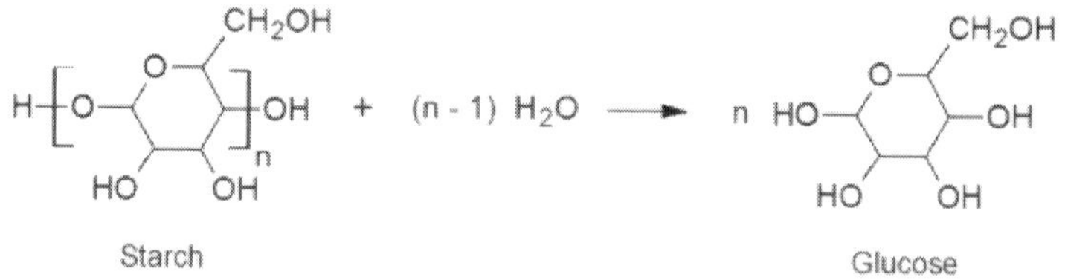

Figure 2.3: Conversion of Starch into Glucose.

Source: https://polymerdatabase.com/polymer%20classes/Polysaccharide%20type.html

The cyclic structure of the starch molecules together with strong hydrogen bonding gives starch a rigid structure and leads to highly ordered crystalline regions. This explains why starch has a high glass transition temperature and melting point and why unmodified starch is only soluble in hot water.

The granules first swell and lose their semi-crystalline structure and then burst. The released amylose and amylopectin molecules gradually dissolve and form a network that holds water. This process is known as starch gelatinization and is the reason why during cooking starch becomes a paste of high viscosity.

Starch is sometimes modified to make it suitable for industrial and food applications. This includes esterification, etherification and oxidation. These chemical modifications are accomplished by the addition of suitable reagents to aqueous starch slurries while controlling the pH and the temperature.

Sodium sulfate or sodium chloride is often added to restrict swelling of the starch granules. After completion of the reaction, the slurry is neutralized with hydrochloric or sulfuric acid, and then filtrated, washed and dried. The degree of substitution of commercial starch is usually rather low but greatly changes its properties.

Depending on the reagents, the reactions lead to nonionic, cationic, anionic or hydrophobic starch which have noticeably different properties. For example, the type and degree of substitution change the gelatinization temperature and the viscoelastic and mechanical properties of starch.

It also affects the stability of the dissolved or dispersed starch granules by controlling or blocking associations of amylose and amylopectin molecules. Certain modifications also improve the freeze-thaw stability which is important for frozen food products.

2.7.1 Starch Esters and Ethers

The two most common starch derivatives are starch acetate prepared by esterification with acetic anhydride and hydroxypropyl starch prepared by etherification with propylene oxide. Esterification is usually carried out at pH 7 - 9 and etherification at pH 11 - 12 and the temperature is usually kept below 60°C. These chemical modifications provide the following benefits:

- They improve the stability
- They lower the slurry/ solution viscosity
- They improve the film forming properties of starch due to increased hydrophobicity

2.7.2 Dextrin and Thinned Starch

Starch is sometimes partially depolymerized which lowers its solution viscosity. This form of starch is often called thinned starch. Sometimes an acid or an oxidant treatment is given which depolymerizes it. This process is called depolymerization.

For example by treating a slurry of granular starch with diluted acetic, hydrochloric or sulfuric acid at 40 - 60°C. Dextrin have an even lower molecular weight. They are produced by exposing dry acidified starch to dry heat.

2.7.3 Polyglucose (Polyglucosides)

Alkyl polyglucose (also called alkyl polyglucosides) like lauryl polyglucose are derived from glucose or starch and fatty alcohols. They are often used as all-purpose fully biodegradable nonionic co-surfactants in (sulfate-free) cosmetics, body washes and shampoos.

2.7.4 Cationic Starches

Quaternary ammonium starch is the most common commercial cationic starch. It is prepared by treating starch with 3-chloro-2-hydroxypropyltrimethylammonium chloride or derivatives thereof under alkaline conditions and at ambient or slightly elevated temperature.

Cationic starches are extensively used as paper additives, emulsion stabilizers, flocculation agents, thickening agents, and sizing agents. Cationic starches can be very fundamental in improving tear strength, can help to bind particles together and on the base paper material and can also increase the retention of fibers and fillers. Therefore, its one of the most important application is the production of paper and cardboard.

2.7.5 Anionic / Oxidized Starch

Commercially important anionic starches include phosphorylated, oxidized and carboxyalkyl starch. The most common form of anionic starch is phosphorylated starch. It is prepared by treating starch with sodium tripolyphosphate. Noteworthy is also carboxymethyl starch, which is prepared by treating starch with sodium mono chloroacetate, and poly (acrylic acid)-starch graft copolymer.

Oxidized or carboxylated starch can be prepared by treating an aqueous or semi-dry starch slurry/paste with hydrogen peroxide under alkaline conditions and at ambient or slightly elevated temperature. This process yields highly *carboxylated starch* or poly (hydroxycarboxylic acids).

Under the right conditions, it is also possible to selectively oxidize hydroxymethyl groups to carboxyl groups (anionic starch). Oxidized starch has improved biodegradability, i.e. it degrades much faster than conventional starch.

Following are the uses of anionic starches:

- They are often used as rheology modifiers
- They are used as thickening agents
- They are used as flocculants and emulsion stabilizers
- They are used as sizing agents
- They can also be used as paper binders
- They are also used as coating agents particularly for food products

2.8 APPLICATIONS OF STARCH-BASED BIODEGRADABLE POLYMERS

It is known fact now that starch can replace petroleum-based polymers with natural ones, so starch can definitely be used to reduce the carbon footprint of traditional resins. It is also highly degradable, meaning it can be used alongside a compostable polymer without interfering with the degradation process.

Starch-based plastics have increased in relevance with the introduction of improved resin grades, their ability to blend with other biopolymers and an increasing number of suppliers. Starch based bioplastics are biocompatible, have low toxicity, degradation properties and mechanical properties. Therefore, these are widely used in the medical industry.

2.8.1 In the Food Industry

Food packaging and edible films are two major applications of the starch-based biodegradable polymers in the food industry. There is a need to reduce the food loss, keep the food fresh, enhance the organoleptic characteristic of food such as appearance, odor, and flavor and provide food safety. Therefore, food packaging is required. Earlier food packaging materials such as LDPE (Low Density Polyethylene) were used which were difficult to dispose of and hence posed an environmental threat.

Figure 2.4: LDPE used for bottle packaging which is not generally environmentally friendly.

Source: https://cdn.pixabay.com/photo/2015/02/07/21/57/plastic-627919_960_720.jpg

The starch-based biodegradable polymers can be a possible alternative for food packaging to overcome these disadvantages and keep the advantages of traditional packaging materials. However, there is a drawback which can be a major concern.

The components in the conventional starch-based polymer packaging materials are not completely inert. The migration of substances into the food possibly happens, and the component that migrates into food may cause harm to the human body.

To overcome this, new starch-based packaging materials are being developed. For instance, a starch/clay nanocomposite food packaging material is developed, which can offer better mechanical property and lower migration of polymer and additives. Starch-based edible films have the following advantages over the traditional ones:

They are odorless, tasteless, colorless, nontoxic and biodegradable

They display very low permeability to oxygen at low relative humidity making them suitable to improve quality and shelf life without hampering consumer's acceptability

In addition to the above advantages, they also have the following benefits:

Starch can be changed into a foamed material by using water steam to replace the polystyrene foam as a packaging material.

It can be pressed into trays or disposable dishes which can dissolve in water and leave a nontoxic solution, that can be consumed by microbes in the environment.

Evidently, the starch-based biodegradable polymers are attractive for the food industry and will make great progress in the future.

2.8.2 In Agriculture

The use of polymers in agriculture is gaining popularity. This has provided solutions to the present-day agriculture related problems which is to maximize land and water productivity without threatening the environment and the natural resources.

Starch-based biodegradable polymers have found three major applications in agriculture: the covering of greenhouse, mulch film and fertilizers-controlled release materials. The consumption of agriculture films is abundant. Usually, the films are disposed of in a landfill, by recycling or by burning.

But these methods are not economic, they are very time consuming and are not environment-friendly. On the other hand, the utilization efficiency of fertilizers is the key element of the development of agricultural productions.

Sometimes fertilizers, due to surface run-off, leaching and vaporization, escape to the environment and cause diseconomy and various environmental problems. The development of starch-based biodegradable polymers offers a possibility to overcome the mentioned problems.

They can be used as the fertilizers-controlled release matrices to release the fertilizers slowly or in a controlled way. As a result, the loss of fertilizers and environmental pollution can be avoided or reduced.

These starch-based films, after their usage, can be disposed of easily by plowing them into the soil. Thus, no toxic leftovers are thereafter its degradation. Thus, the development of starch-based materials for agriculture applications is being continued.

For example, to enhance the mechanical properties and solvent or gas resistance, starch-based biodegradable materials are mixed with some Nano-grade additives such as TiO_2, layered silicate and MMT to form bionanocomposites.

2.8.3 In the Medical Field

Starch-based biodegradable polymers have proved to be good medical polymer materials because of the following advantages:

- good biocompatibility
- biodegradable and its degradation products are non-toxic
- proper mechanical properties
- degradation as requirements

Several investigations have been carried out to use starch-based biodegradable polymers in bone tissue engineering. They have proved to be useful because of the following:

Starch-based biodegradable bone cement can provide immediate structural support and degrade from the site of application.

Moreover, they can be combined with bioactive particles, which allow new bone growth to be induced in both the interface of cement-bone and the volume left by polymer degradation

Starch-based biodegradable polymers can also be used as bone tissue engineering scaffold.

Starch-based biodegradable polymers, in the form of microsphere or hydrogel, are suitable for drug delivery. There is no need for surgical removal of the device after drug depletion.

Starch-based biodegradable polymers have hydrogels which make them useful for various biomedical applications because of their hydrophilicity, permeability, biocompatibility, and similarity to soft biological systems. The starch-based hydrogels are able to absorb and reserve plenty of water and also keep good enough mechanical property at the same time because of their 3D structure. Starch-based hydrogels have received growing interest for biomedical applications.

In our lab, physically cross-linked starch-gPVA and starch-g-PVA/hydroxyapatite hydrogel are obtained via repeated freezing/thawing circles, and hydroxyapatite (HA) can be well dispersed in such a matrix. The water content in the fresh starch-g-PVA/HA hydrogel is comparable to that of PVA/HA hydrogel, and the dried starch-g-PVA/HA films can re-absorb water soon and reach swelling equilibrium within 12 minutes.

2.8.4 Starch Based Films

Starch-based films found in the market are made mainly from starch completed with thermoplastic polyesters, such as poly-epsilon caprolactone (PCL), to form biodegradable and compostable products (Bastioli, 2000).

When used for the manufacturing of bags for the recycling of organic waste (composting), packaging, hygienic items and agriculture, the properties of these films are similar to those of LDPE (Low Density Polyethylene).

Supermarkets in Scandinavia and in the Mediterranean Coast were among the firsts to introduce Purchase Bags (Bastioli, 2000) in 1999. They were introduced in places where the collecting of organic wastes already existed and where they were accepted as biodegradable compost bags.

These bags have a behavior similar to the bags made with LDPE as regards the maximum weight and most of its performances.

Consumer goods packaging (Bastioli, 2000): There are markets for magazine wrapping and bubble films mainly for electronic goods besides the main market being of silk paper.

Food packaging (Bastioli, 2000): There is a need for the bags of fruits, vegetables and bakery products. The cost of the biodegradable products is a big limitation here but since these starch films allow for a better breathing of the products, they are still used over traditional plastics.

Composting bags (Bastioli,2000): Bags used in the selective collecting of organic waste, which will be treated to produce a compound. Millions of Europeans already use these bags.

Hygiene–cosmetics (Thouzeau,2001): These bags are used in the hygiene line of business. These biodegradable polymers are found in diapers, swabs and even toothpicks.

Funerary goods (Thouzeau, 2001): Biodegradable plastics can be used to wrap corpses, in compliance with the rules on the use of biodegradable materials.

According to Tuiletal (2000), multilayered biodegradable films can be used as food packaging with either controlled atmosphere or gas. Starch packagings is an excellent barrier to oxygen and carbon dioxide. But they face problems with water and water vapor.

The application of layers of biopolymers resistant to water in the ratio of 10% allowed Tuiletal. (2000) to obtain films with a thickness from 30 to 100 mm that are resistant to water. This material is currently commercialized by AVEBE under the name of Paragon® and can be processed by means of blowing and used in multilayers with PLA or PHBV.

2.8.5 Starch Based Expanded Packaging

Two main markets for starch derivatives expanded products:

Loose fill products: This is the main market of biopolymers. The average density of snack-type products is 8kg/ cubic meter, compared to 4 for EPS derivatives (expanded polystyrene). The main market is the United States.

Disposable food service ware (Bastioli, 2000): Many governments had to either ban the use of EPS or had to start levying taxes on them because they were causing a lot of environmental problems. Starch or polylactic-based extruded or laminated products are available in the market, and they are only slightly more expensive compared to traditional products.

Thermoformed PLA products are also available. Products using bakery technology, following the principle of wafers production, are also found in the market, but only on an experimental basis. Tuilet al. (2000) affirm that expanded packaging made by means of bakery technology, or wafer technology, are about to enter the market.

Technologies developed and available in AVEBE allow the development of trays-type packaging with properties similar to those of XPS (Extruded polystyrene) and EPS, biodegradable and resistant to moisture. Products expanded entirely from PLA are still under study (Weber,2000)

It is believed that the blend of starch with PVOH (polyvinyl alcohol) can be helpful for the production of materials to replace expanded polystyrenes, so various studies have been conducted in this regard.

According to Cha et al. (2001), a blend with 95% of hydroxy propylated corn starch with a high content of amylase and 5% of PVOH is used as a commercial alternative to the poly- styrene. For Glennetal (2001-b), a blend of starch with PVOH, expanded by means of wafer technology, leads to a decrease from 25 to 32% in water absorption.

According to the authors, cross-linked agents decrease this absorption by 80%. Chandra and Rustgi (1998) affirmed that the isocyanide is a good cross-linking agent.

2.8.6 Blend of Starch with Cellulose Derivatives

The mechanical properties, permeability to gas and resistance to water (Averouset al 2001 b) can be improved by the incorporation of cellulosic fibers to TPS because cellulose is less hydrophilic than starch. Cellulose-starch interactions mobilize sites previously used by water molecules (Averouset al. 2001-b).

Nevertheless, the incorporation of fibers is limited due to the difficulties of dispersion (Dufresne et al.,1997.). According to the authors, the presence of pectin increases the cohesion of fibers with the material, increasing pressure forces, mainly in environments with low humidity. With the increase in relative humidity, pectin, which is more hydrophilic, absorbs a larger quantity of water and gets softer, impairing the material's mechanical properties.

According to Glenn et al.(2001 a), wood fibers with a length from 5 to 7 mm greatly improve the flexibility and other functional properties of starch derivatives processed by means of wafer technology. The production of vegetal polymers-based composite materials is still in the early stages of development. It consists of formulations associating vegetal polymers to a linking or thermoplastic matrix (starch, resin) and several adjuvant elements.

It is possible to mention products, like FASAL and NAPAC which are appearing in Germany, Switzerland or Austria, for the production of parts molded by thermopressing or injection processes. The polymer made of triglycerides and polycarbonate acid anhydride, developed by PREFORM in Germany, is used as a linkage for fibers (wood, straw etc.) in the production of extruded parts or formed panels.

In Brazil, little pots made with fibers and a starch-based link is being tested for eucalyptus and pines plants production. According to Arvanitoyannis and Biliaderis (1999), methylcellulose provides a homogeneous dispersion with the starch and it is less hydrophilic, allowing the starch to enhance its weak barrier to moisture. For the authors, the hydrophilic character of the blend increases with a higher content of plasticizing elements (water or a blend of water-sugar-glycerol). However, despite the enhancements, films are still not very resistant to moisture.

2.9 STARCH-FILLED PLASTICS OPTIONS FOR AUTOMATION

Two new grades of polypropylene incorporating encapsulating starch particles will be released to the market by Cereplast, Inc. (El Segundo, CA) during the first half of 2012 in the United States and Europe.

Hybrid 102D is a formulation that offers higher starch content than the current Hybrid 101 grade while maintaining similar performance. It is primarily used in injection molding which requires some ductility.

Hybrid 105D is a high flow injection molding grade for thinner wall applications than the current Hybrid 101 offering and replaces Hybrid 103, providing more consistent properties and processing. Cereplast Hybrid Resins are suitable for durable goods, including consumer products, interior automotive parts, and furniture, and have a lower carbon footprint than traditional plastics.

"As demand for bioplastic resins grows, so does the demand for high-performance durable grades that are cost effective," says Chairman and CEO of Cereplast, Frederic Scheer. "As the price of oil continues to rise, Cereplast Hybrid Resins get closer to parity with traditional plastics, providing an affordable, eco-friendly solution for companies that wish to reduce their environmental impact.

Our research and development team have answered the demands of our clients for better-performing hybrid resins with higher levels of starch content. We are proud to introduce the next generation of bioplastic hybrid resins." Cereplast Hybrid Resins replace up to 50% of the petroleum content in traditional plastic products with bio-based materials such as starches from annually renewable plants.

2.10 INFLUENCE OF STARCH OXIDIZATION AND MODIFICATION ON INTERFACIAL INTERACTION, RHEOLOGICAL BEHAVIOR, AND PROPERTIES OF POLY (PROPYLENE CARBONATE)/STARCH BLENDS

The enhanced maleic anhydride-end-capped poly (propylene carbonate)/ starch blends were prepared through starch oxidization and modification with a coupling agent, aluminic ester. The interfacial interaction, rheological behavior, and properties of blends were investigated through Fourier transform infrared spectroscopy, rheological measurement, mechanical property test, differential scanning calorimetric, thermogravimetric analysis, and moisture absorption test.

The results show that hydrogen-bonding interaction is formed between poly (propylene carbonate) and starch, which makes the tensile strength of maleic anhydride-end-capped poly (propylene carbonate)/starch blends improve significantly.

The glass transition temperature (T_g) of blends is increased when a coupling agent is induced into the polymer system. When increasing the content of starch modified with a coupling agent from 10 to 30%, T_g values for composites increased from 30.5 to 32.8°C.

Thermogravimetric analysis results show that oxidation of starch can improve the thermal stability and modification of starch through aluminic ester that can further increase the thermal stability of maleic anhydride-end-capped poly (propylene carbonate)/starch blends. Oxidation of starch has no significant effect on moisture absorption for poly (propylene carbonate)/ starch blends.

2.11 STARCH-BASED PLASTIC BLENDS FROM REACTIVE EXTRUSION

The North American market for biodegradable plastics in 2005 was estimated to be around 60 million pounds (27 kt). Starch-based polymer blends were expected to account for 30 million pounds (14 kt) with significantly lower growth rates than other biodegradable polymers such as polyesters. The main hurdle in the growth of starch-based products is the thermodynamic immiscibility and non-wetting of starch with other polymers which leads to serious deterioration of mechanical properties at >25–30 wt% starch.

Higher amounts of starch in the blends entail adding suitable functional groups on starch and other polymers in the blend to make them more compatible. The primary challenge is to develop fast reaction chemistries that can be transformed into viable processes and integrated into existing process lines with economically viable formulations. This article briefly reviews some of the most promising chemistries available for the reactive extrusion of starch-based polymer blends (biodegradable/non-biodegradable).

2.12 STARCH FILLED PLASTICS: THE FUTURE OF SUSTAINABLE PACKAGING

2.12.1 Starch Blends with Compostable Polymers

Starch-based plastics are complex blends of starch with compostable plastics such as PLA, PBAT, PBS, PCL and PHAs. If the starch is blended with plastics, it is noticed that the water resistance, processing properties and mechanical properties are improved considerably.

Starch based trays are not transparent. Other packaging products Starch-based materials are frequently used in loose fill foams for transport packaging. Another application is in service ware like cups, plates and cutlery.

Biodegradable films, with starch as a matrix, were developed and reinforced with wheat and corn hulls. It was observed that the addition of hulls enhanced the modulus, tensile strength, and impact strength of the starch matrix at the expense of its elongation. The water-vapor transmission rate results show that corn starch was more efficient in reducing the water-vapor permeability than wheat hulls.

The challenges for researchers and the packaging industry in terms of producing starch-based blends with commercial utility are:

- Overcoming miscibility problems at high starch contents,
- Avoiding mechanical property deterioration at high starch contents, even in compatibilized blends,
- Reducing costs, especially for biodegradable starch-polyester blends at low starch contents (<30 wt. %)

2.12.2 Antimicrobial Packaging Film

It is very important that the food products are free from any kind of microbial growth and have a longer shelf life. In order to achieve this, an antimicrobial agent is integrated into packaging systems. This process is called antimicrobial packaging.

- The packaging materials may acquire antimicrobial activity
- By incorporating antimicrobial components in a polymer matrix,
- Surface irradiation of polymer matrix which produces reactive oxidizing species,
- By gas emission/flush through modified atmosphere packaging,

By using inherently antimicrobial polymer resins. The use of antimicrobials in food packaging systems has been motivated with the increasingly global food-borne outbreaks with relation to health and safety concerns.

Antimicrobial packaging has two main categories, namely migratory and non-migratory packaging systems.

- Migratory packaging system allows the reversible release of non-volatile or volatile active components from polymer matrix to food constituents or packages headspace by diffusion and/or partition at interfaces.
- In the non-migratory system, the active component is irreversibly tethered to the package's surface and no diffusivity of the antimicrobial agents occur.

These additives can be deactivated in the food matrix or migrate out of the food surface towards locations in the food where initial microbial attacks are not taking place. The natural antimicrobial agents have broad spectrum actions against most of pathogenic and food spoilage microorganisms.

Additionally, nanomaterials from titanium, magnesium, copper, silver, platinum, gold and zinc have also received attention in antimicrobial food packaging due to their large surface-volume ratio. Ternary blend films were prepared with different ratios of starch-polyvinyl alcohol with citric acid to obtain films with better antibacterial, mechanical, and thermal properties.

In other work two essential oils, Zataria multiflora Boiss (ZEO) or Mentha pulegium (MEO) at three levels (1%, 2% and 3% (v/v)), were incorporated into starch films using a solution casting method to improve the mechanical and water vapor permeability (WVP) properties and to impart antimicrobial activity. Plants contain polyphenolic compounds and a large number of them were used as antioxidant, antibacterial and antifungal properties.

2.12.3 Starch-based Nanocomposite Films

Natural materials such as clay nanofillers with starch used to develop biodegradable nanocomposite packaging films to address issues of environmental impact t, agriculture sustainability. 7 - 9 parts of Nano-Montmorillonite in a plastic film has shown high strength. The prepared film has a lot of benefits which are as follows:

- It can be completely degraded.
- Even after it is discarded, it is environment-friendly.

- It can be prepared into plastic bags, preservative films and the like for food packages.
- It is worthy of large popularization.

It was also noticed that after the addition of light calcium carbonate and rock powder, the qualities of finished products were improved, their cost was reduced by 45% and it had a good degradation effect.

Nano silica was also added. It improved the mechanical properties of the nanocomposite sheets. It had the following favorable effects:

- Its tensile strength was increased. It also adversely affected the elongation at break.
- The sheets also displayed improved water resistance while those without Nano silica disintegrated within two hours of immersion in distilled water at 25

Food packaging was originally used to protect food against heat, light, moisture, oxygen, microbial attack, insects and other impurities. In one study, a corn starch-based nanocomposite sheet was prepared using Nano silica/ from rice hull ash to enhance mechanical and water absorption packaging properties of corn starch.

Cellulose nanofillers could be considered as environmentally friendly and renewable nanofillers. Nanocrystalline cellulose (NCC) from sugar palm fibers and nanocrystals from sugar palm starch were prospective Nano-reinforcement or Nano-filler to enhance the properties of sugar palm starch-based films as high-performance packaging material. Nano-clays have been added to various polymeric matrices in very small quantities to improve film properties.

It has been proved that Nano fillers induce higher mechanical strength and improve barrier against gases and water vapor. These fillers are capable of improving barrier and mechanical properties by decreasing filler dimensions and also reducing production cost due to lower material consumption. Moreover, Nano-filler reinforcement affects polymer crystallinity, reduces transport and enhanced water absorption.

Nano clay is also added to starch composite films because it has been found that it improves the mechanical strength and transport properties. Microporous, aluminosilicate minerals like Zeolites can also use as fillers in starch-based films. Studies show that Zeolites leads to the enhancement of Young's modulus and reduction of gas and vapor transport and also water solubility.

Nano-cellulose is also added to starch in order to enhance starch film properties. Based on the results, Nano-cellulose leads to the improvement of mechanical properties up to 70%. NCC was used as a Nano-scale additive into the starch chitosan and gelatin-chitosan Nano-composite films. Chitosan was used to enhance antibacterial and anti-fungal properties.

Many investigations were developed on the natural fiber's potential as reinforcements for composites. Starch and poly (vinyl alcohol) (PVOH)-based packaging is of great importance and had additional benefits over petroleum-based packaging due to abundant availability, biodegradability and compatibility.

Presence of laponite RD improved the mechanical barrier properties of starch/PVOH matrix up to10% level of laponite RD because of better surface interface interaction between polymer matrix and layers of laponite RD.

Film-forming formulations comprising starch, ethylene acrylic acid copolymer, and optionally polyethylene, can be blown into films upon neutralization of a portion of the copolymer acid functionality. The resultant biodegradable films had potential application as agricultural mulch, garbage bags, and various types of packaging.

The elaboration of flexible films from cassava starch for the manufacture of biodegradable packaging useful in the packing and packaging of dry foods and other products. The novel films were produced by extrusion of a mixture of cassava starch and plasticizer. A series of corn starch films with varying concentrations (0% - 20%, W/W) of citric acid (CA) and Carboxymethyl cellulose (CMC) were produced by the casting method.

As a result of its multi-carboxylic structure, the interaction could take place between the carboxyl groups of CA and the hydroxyl groups on the starch. Such an interaction would improve the water resistibility due to reducing available OH groups of starch.

Cellulose-based fibers are widely used as biodegradable filler. When natural fibers were mixed with starch, the mechanical properties of the resulted composite were obviously improved, because the chemical similarities of starch and plant fibers provide a good interaction. There are few papers about the CMC/starch biocomposite properties. The addition of CMC improved the moisture resistance of the composites.

2.12.4 Heat Sealing Packaging

Heat sealing capacity of native and acetylated corn starch-based films was evaluated to develop biodegradable packages, such as bags. Acetylated starch addition reinforced 80% sealing resistance of starch films. Numerous sealing techniques are available.

The mechanical and barrier properties of amorphous polymers are determined, and the crystallization kinetics of these materials is controlled by the glass transition temperature (Tg).

Acetylated starch addition in film formulation reinforced the sealing resistance. Heat sealing tests were conducted. Following results were found:

- Un-plasticized films exhibit adhesive failures.
- Those containing glycerol presented a rupture near the zip.

Rice that is broken during processing is used to make rice flour which is starchy, and it is low cost too. Films from rice starch and rice flour were prepared by casting, with glycerol or sorbitol as a plasticizer. Rice flour films prepared had similar mechanical properties to those of starch-based films. However, their water vapor permeabilities were two times higher than those of starch-based films. The production of biodegradable and edible films from carbohydrates and proteins adds value to low-cost raw materials and can play an important role in food preservation.

As the glycerol content in the potato starch films is increased, its macromolecular mobility also increases. Also, potato starch films become less stiff and more flexible. The transparency was found to decrease. Active and smart biodegradable films from cassava starch and glycerol with 5 wt.% of different natural extracts such as green tea and basil were prepared and tested for degradation.

These films degraded in soil under two weeks and were thermally stable up to 240°C. The incorporation of extracts of green tea and basil was found to make films with lower water vapor permeability retaining their flexibility.

2.12.5 Starch Grafted Emulsions for Packaging

The physical and chemical properties of polymers can be modified by using a technique called grafting. The graft copolymerization of synthetic polymers onto a starch backbone was one of the best ways of improving the properties of starch. Starch grafted on vinyl emulsion also undergoes biodegradation and can be used as sustainable packaging material.

REFERENCES

1. Abeke Omotoso, M., Ayorinde, G. and Alex Akinsanoye, O. (n.d.). *Preparation of Biodegradable Plastic Films from Tuber and Root Starches*. [ebook] Available at: http://www.iosrjournals.org/iosr-jac/papers/vol8-issue4/Version-1/C08411020.pdf [Accessed 28 Mar. 2019].
2. Bastioli, C. (1995). Starch-polymer composites. *Degradable Polymers*, [online] pp.112-137. Available at: https://link.springer.com/chapter/10.1007/978-94-011-0571-2_6 [Accessed 28 Mar. 2019].
3. Gadhave, R., Das, A., Mahanwar, P. and Gadekar, P. (2018). Starch Based Bio-Plastics: The Future of Sustainable Packaging. *Open Journal of Polymer Chemistry*, [online] 08(02), pp.21-33. Available at: http://file.scirp.org/pdf/OJPChem_2018052915415893.pdf [Accessed 28 Mar. 2019].
4. Green Dot Bioplastics. (2019). *The utility of starch-based plastics - Green Dot Bioplastics*. [online] Available at: https://www.greendotbioplastics.com/starch-based-plastics/ [Accessed 28 Mar. 2019].
5. Jiang, G., Xu, J., Zhao, N., Zhang, S. and Huang, H. (2016). Influence of Starch Oxidization and Modification on Interfacial Interaction, Rheological Behavior, and Properties of Poly(Propylene Carbonate)/Starch Blends. *Polymer-Plastics Technology and Engineering*, [online] 56(10), pp.1084-1095. Available at: https://www.tandfonline.com/doi/abs/10.1080/03602559.2016.1253736?scroll=top&needAccess=true&journalCode=lpte20 [Accessed 28 Mar. 2019].
6. Kalambur, S. and Rizvi, S. (2006). An Overview of Starch-Based Plastic Blends from Reactive Extrusion. *Journal of Plastic Film & Sheeting*, [online] 22(1), pp.39-58. Available at: https://journals.sagepub.com/doi/10.1177/8756087906062729 [Accessed 28 Mar. 2019].
7. Kumar, D. (2017). *Bio-Plastics From Starch Based Food Crops*. [online] Available at: https://www.ramauniversity.ac.in/blog/bio-plastics-from-starch-based-food-crops/ [Accessed 28 Mar. 2019].
8. Lu, D., Xiao, C. and Xu, S. (2009). Starch-based completely biodegradable polymer materials. *Express Polymer Letters*, [online] 3(6), pp.366-375. Available at: http://www.expresspolymlett.com/letolt.php?file=EPL-0000946&mi=dc [Accessed 28 Mar. 2019].
9. PlasticsToday. (2012). *More starch-filled PP options for auto use.*

[online] Available at: https://www.plasticstoday.com/content/more-starch-filled-pp-options-auto-use/40509108116952 [Accessed 28 Mar. 2019].

10. Polymerdatabase.com. (2019). *MODIFIED STARCH & STARCH-BASED PLASTICS*. [online] Available at: https://polymerdatabase.com/polymer%20classes/Polysaccharide%20type.html [Accessed 28 Mar. 2019].

11. Vilpoux, O. and Averous, L. (n.d.). *Starch-Based Plastics*. [ebook] Available at: http://www.biodeg.net/fichiers/Starch-based_Plastics.pdf [Accessed 28 Mar. 2019].

Chapter 3

CELLULOSE AND CELLULOSE ACETATE: BIO-BASED PLASTICS

CONTENTS

Cellulose is one of the most abundant natural polymer present on the earth. It constitutes an inexhaustible source of the raw material so as to produce biocompatible and environment friendly products. Cellulose does not melt because of an extended network of hydrogen bonds. It cannot be easily processed like other conventional thermoplastics. At the beginning of this chapter, an overview of cellulose fibers is given. After that, the degradation process of cellulose acetate and synthesis of cellulose-based plastics has been described. At the end, the chapter deals with the requirement of sustainable raw material for cellulose-based plastics obtained from healthcare sectors, food and ceramics.

3.1 INTRODUCTION

Materials made of plastics have a number of applications in day-to-day life. Commonly used plastics are presently made using fossil fuels like petroleum, etc. and used and dumped in the environment. The degradation of these plastics is a time-consuming process and thus causes the problems related to their collection and dumping, which generally leads to filling of landfills or in waste incineration for energy recovery.

The consumers and new environmental and waste management policies are pressuring the manufacturers, including the increased oil prices, to shift from the polymer technology used for the manufacture of traditional material to the materials that are produced b renewable resources.

Cellulose is a natural polymer found in abundance on earth, comprising an infinite source of raw material used for producing biocompatible and environment-friendly products that need not compete with the food chain.

Like traditional thermoplastics, cellulose cannot be processed. It can be processed using chemical modification. This is because of the presence of an extended hydrogen-bond network. In this chapter, an overview of cellulose raw materials and its molecular and supermolecular structure is given.

Further, the principles of shaping cellulose into films, fibers, and nonwovens using solution techniques are highlighted, tracked by a segment on the market use and properties of these materials. Derivatives of cellulose are depicted, especially emphasizing on typical plasticizers, cellulose esters, and promising reinforcing materials.

The current developments and upcoming visions are finally reviewed for cellulose materials concerning the discussed applications discussed below.

It is commonly known that a suitable raw material for thermoplastic processing is secondary cellulose acetate (with 53 to 56 percent acetyl groups). In order to obtain a plastic material which outstands in pleasant texture and transparency, suitable plasticizers are added. The plastic materials thus obtained have various applications like in spectacle frames, tool handle, combs and many more. In principle, cellulose acetate with such a degree of substitution is biodegradable, although degradation proceeds extremely slowly.

When certain esters and low molecular components (at least 30 percent by weight) are replaced some common plasticizers, the "plastic material" possess the same properties as of thermoplastics but will decompose in water or soil within a short span of time, maybe just a few years

For instance, a cup made of cellulose acetate when buried in sewage sludge lost its original weight by more than 70 percent within a time period of about eighteen months. In an aqueous environment (Sturm's test), the decomposition amounted to more than 60 percent. The process of degradation is faster under favorable conditions of composting or conditions favoring anaerobic (in the absence/limited supply of oxygen) fermentation. Apart from decomposition, the material can also be incinerated without any leftovers or recycled.

These new age biodegradable plastics can be processed using conventional injections molding machines or on extruders adapting their certain processing properties. They can be changed into strips, granular materials and films and into extrusion blow-molded shapes and injection molded.

The properties of materials are comparable to the properties of known cellulose ester-based plastics. Films and thin-walled hollow bodies are quite transparent. The mechanical properties are quite comparable with the properties of traditional packaging materials.

In comparison to standard plastics, these materials have low thermal resistance and high permeability to oxygen and steam. The material is not affected by oils and fats. Also, it is resistant to acids and alkalis for short period of time.

The esterification of cellulose produces a thermoplastic material, Cellulose acetate (CDA) which is an utmost important cellulose-derived biopolymer as per the industrial viewpoint. Cellulose esters can be produced in powder form by using a number of raw materials, like wood cellulose, cotton, sugarcane and recycled paper.

It has been stated that cellulose acetate is possibly biodegradable. Thus, fresh applications of CDA can be foreseen not only in the sectors of packaging, electronics industries and automotive but also in pharmaceutical and medical industry, biocompatibility assessment of full blends in the production of biocomposite materials and for other bio-related fields.

The glass transition temperature (T_g) of cellulose acetate is high. Thus, it cannot be melt-processed as a raw material as it begins decomposing before melting. Hence, plasticizers are used in CDA to decrease its glass transition temperature (T_g) and processing temperature.

Conventionally the plasticization of CDA is accomplished using glycerol derivatives, citrates, phosphates, phthalates, etc. From earlier times the most common plasticizers used industrially for CDA is Phthalate esters. These Phthalate esters, when subjected to environmental examination, came out as a threat to health and thus their long-term application is a serious matter of concern. For improving the processing of CDA, new plasticizers have been explored, for instance, propylene glycol, poly (caprolacton triol), and polyethylene glycol.

Others considered maleic anhydride, glycol and triacetin (TA) as multi-plasticizers. The research activities upgraded the processing of CDA based materials but at the cost of their mechanical properties. According to their research and examination, Diacetin (DA) and Triacetin (TA) were selected to enhance CDA processing. The reason for this is that these plasticizers are less toxic and biodegrade fast, thus becoming environmentally friendly and sustainable.

As discussed before cellulose decomposes before melting but cannot be plasticized or dissolved with common plasticizers or solvents. This is due to its high degree of crystallinity, making it more stable as compared to starch. It is must to chemically modify the plastic material for the production of cellulose in the form of pulp. "The most important thermoplastic cellulose material is cellulose acetate (CA) (Balser et al., 2000)."

CA remained the chief plastic material from the time of its first commercial production at the beginning of the twentieth century. But now it is being substituted by petroleum-based plastics bearing low costs. The process of acetylation of the hydroxyl group of pulp with acetic anhydride results in the formation of Cellulose Acetate (CA).

The CA is partially or completely bio-based is decided by the source of the anhydride. "The thermal and mechanical properties and the biodegradation rate of CA depend on the degree of acetylation". (Chandra and Rustgi, 1998).

CA is extensively used in fiber applications like in cigarette filters and textiles. Niche plastic applications account for handles and casings of other high-value products, polarized films, which are used for LCD displays and spectacle frames.

Deriving pulp industrially is a sophisticated and developed process that is conducted under severe conditions. "Recent research has focused on alternative solvents, such as ionic liquids." (Gericke et al., 2012). But this technique has a major limitation that the cost of the recovery of these solvents is quite high. The main producer of cellulosic polymers is Eastman Chemical (USA), followed by FKuR (Germany), Innovia Films Ltd (UK).

Plastics applications require only a small part of CA production (800 ktpa). It is difficult to expect any major changes in the production capacities because of the high cost and the niche character of most applications.

3.2 CELLULOSE FIBERS

Cellulose fibers are being used extensively because of its high beneficial values. These beneficial values or advantages include low weight, exhibit good mechanical properties, biodegradable, abundantly available, low abrasive nature, interesting specific properties, cheaper and since these are waste biomass.

In spite of so many advantages, there are some drawbacks of cellulose fiber also. These disadvantages include poor compatibility with the hydrophobic polymer matrix, moisture absorption, low thermal stability and quality variations.

3.2.1 Chemistry of Cellulose

One of the most abundant forms of living terrestrial biomass is Cellulose, which is applicable in number sectors in the current industry. Anselm Payen in 1838 was the first to identify the presence of cellulose as a common material in the cell wall of plants.

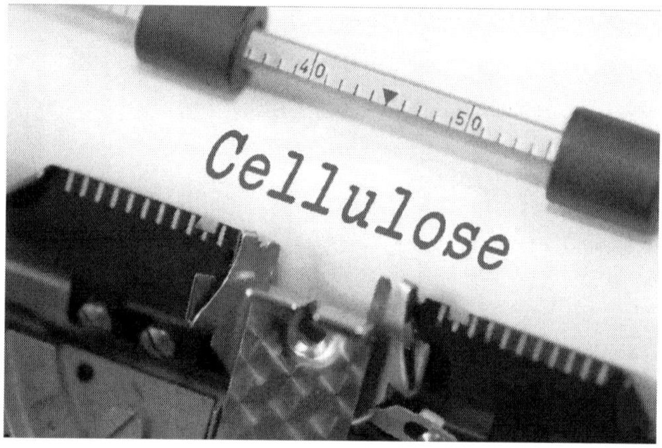

Figure 3.1: Cellulose.

Source: http://www.thebluediamondgallery.com/typewriter/images/cellulose.jpg

Cellulose represents a long-chain polymer constituting repeating units of simple sugar, D-glucose. It is found in cotton fibers in pure form. Whereas, in plant leaves, stalks and wood, it occurs in a combined state with some other materials like hemicelluloses and lignin. Even though it is considered a plant material, but it can also be prepared by some bacteria.

The natural polymer, cellulose comprises a long chain linked by smaller molecules. The links in the cellulose chain consist of sugar, β-D-glucose. These sugar units get linked when water is removed by the combination of H and –OH group.

A disaccharide termed cellobiose is produced as a result of the linking of just two of these sugars. In the cellulose chain, the glucose units are in 6-membered rings, called pyranoses. They are joined by single oxygen atoms (acetal linkages) between the C-1 of one pyranose ring and the C-4 of the next ring.

Anhydrous glucose units are the glucose units in cellulose polymer that are created with the loss of a molecule of water due to the reaction of hemiacetal and alcohol in order to form an acetal. It is very important to understand the stereochemistry or the spatial arrangement of these acetal linkages. In layman terms, it can be understood as the linkage is in equatorial position or it can be said that pyranose rings of cellulose molecule are sticking out of the periphery of rings due to the fact that pyranose rings have all groups larger than hydrogen.

Further, talking of stereochemistry of pyranose, it can be seen that carbons at positions 2, 3, 4, and 5 of the glucose molecule are fixed, in case of pyranose, the carbonyl at position C-1 can be approached by hydroxyl at position C-4 from any of the sides, and as a result, two different types of stereochemistry is obtained at position C-1.

The two stereochemistry are named as α configuration and β configuration. α configuration is achieved when the hydroxyl group at C-1 is found to be on the same side as the C-6 carbon, and if the C-1 oxygen is on the opposite direction in cellulose, the configuration obtained is a β configuration or in nomenclature terms, it can be said as poly[β-1,4-D-anhydroglucopyranose]. When all the functional groups are in the equatorial position in β configuration, it allows the cellulose molecular chain to grow in approximately in a straight-line fashion, making this configuration a polymer good for forming fibers.

On the cellulose chain, due to the equatorial positions of the hydroxyls, hydrogen bonding is easily achieved due to the fact that hydroxyls bulge out laterally along the extended molecule and are easily available. A highly ordered structure is obtained through these hydrogen bonds as these chains club together. Due to the fact that these chains are commonly longer as compare to the crystalline regions, it is possible for them to pass through the various and diverse crystalline structure, having the disordered area inside the structure mimicking the "fringed-micelle" model. The interchain hydrogen bonds present in crystalline parts are considered to be solid and tough, resulting in insolubility (in most of the solvents) and good strength to the fiber obtained.

These hydrogen bonds further inhibit the cellulose from melting and these celluloses are hence termed as non-thermoplastics also. In the regions that are not in proper order, it has been observed that chains are present at a higher distance and are more active for hydrogen bonding with different molecules present, like water. Most of the cellulose structures are hygroscopic that they can absorb an enormous amount of water. And as a result, these cellulose fibers swell up but does not dissolve in water. There are three different types of hydro glucose units:

- the non-reducing end with a free hydroxyl at C-4
- the reducing end with a free hemiacetal (or aldehyde) group at C-1
- the internal rings joined at C-1 and C-4

But due to the long length of chain, the reactions and chemistry predominate the alcohol groups present in the internal unit, as long as chains are not sliced due to the conditions of the reaction. Nevertheless, as compared to simple alcohols, steric factors are the ones that usually control the cellulose reactions. The role of steric factors is described by different hydroxyl groups and their inherent reactivity.

C-H groups along with C-2, C-3, and C-6 hydroxyls are active sites in cellulose for the amalgamation of polymeric chains with the help of grafting. In the process of grafting, it has been seen that the reactivity of the hydroxyl group at C-2 and C-3 is far more than those at C-6.

3.2.2 Chemical Composition, Structure, and Properties of Cellulose Fibers

Cellulose fibers can be differentiated as per their origin and are categorized into

- Leaf: cantala, abaca, date palm, curaua, pineapple, henequen, banana, sisal
- Bast: jute, hemp, ramie, flax
- Seed: Cotton
- Fruit: oil palm, kapok, coir
- Grass: bagasse, alfa, bamboo
- Stalk: straw (cereal)

The types of hard fibers like leaf and Bast are generally used in composite applications. Some of the commonly used plant fibers are jute, flax, cotton, hemp, ramie, coir, sisal, flax, kapok and henequen. Tanzania and Brazil are the largest producers of sisal around the globe. Mexico is the biggest producer of Henequen whereas hemp and abaca are majorly produced in the Philippines. The largest producers of jute are China, India, and Bangladesh.

The major constituents of cellulose fibers majorly consist of microfibrils that are cellulose which is bound in a helical manner and linked together with the help of amorphous lignin matrix. The main function of lignin is to keep the water intact within the fibers, and acts as a protective shield against the biological attack and also functions as a stiffener, so as to provide stem with resistance against the forces like wind and gravity.

Figure 3.2: Henequen is found in Mexico.

Source: https://c1.staticflickr.com/6/5774/20879421524_316b7dced7_b.jpg

Lignin and cellulose are believed to be compatibilized with hemicellulose that is found in natural fibers. The cell wall found in fiber is a heterogeneous membrane. Each fiber has a layered and complex structure that consists of a thin primary wall which is a first layer collected at the time of cell growth that encircles a secondary wall. There are three layers in the secondary wall and the thick middle layer is used to determine the mechanical properties of the fiber. The thick central layer consists of a series of helical wound cellular microfibrils that are formed with the help of long-chain cellulose molecules. The angle between the microfibrils and the fiber axis is called the microfibrils angle.

The typical value of microfibrillar angle differs from fiber to fiber. The typical diameter of microfibrils is about 10-30nm and are formed by 30–100 cellulose molecules in extended chain conformation and give mechanical strength to the fiber.

Many factors alter the properties of cellulose. This disintegration (mechanical, steam explosion treatment), retting degree, technical processes (carding and spinning), fiber modification, decortications, climate, maturity, variety, textile and harvest.

It is important to understand the physical, chemical and mechanical properties of natural fibers, for understanding the properties of natural fiber-reinforced composite materials. In comparison to other natural fibers, flax fibers are strong. The tensile strength or the extent of the stretch ability of the elementary fibers lies in the range of 1500 MPa and for technical fibers, a value of circa 800 MPa was observed at 3 mm clamp length. Lamy and Baley examined the modulus of flax fibers.

The modulus of elementary fibers is dependent on the diameter of the fiber and it ranges from 39 GPa for fibers having a diameter approximately 35 μm to 78 GPa for fibers having 5 μm diameter. The difference in lumen size between fibers with varying diameter is related to the above variation.

The approximate value of Young's modulus of 54 GPa was seen after performing a number of tensile tests on single flax fibers and the results lie somewhere between the region of moduli measured in technical fibers. The physical, mechanical and chemical properties of plant fibers depend on certain factors strongly, like weather conditions, climate, harvest, soil characteristics. The processing of fibers like scotching, spinning, retting, bleaching also affects the aforesaid properties.

Cellulose fibers have comparatively low density, high strength and high stiffness. The value of E-glass fibers is approximately observed to be close to the specific values of flax and soft-wood-Kraft fibers. While the processing period is on, the natural fibers can incorporate a number of mechanical properties.

The structure and properties of fiber depend on various conditions. They alter with changes like age of the plant, area of growth and climate. One more major factor that helps in determining the specific value and structure of the fiber, is Technical digestion.

Nearly 10 GPa is the elastic modulus of the bulk natural fibers like wood. Chemical-pulping process is used to separate cellulose fibers with moduli up to 40 GPa from wood. These type of fibers can be further subdivided into microfibrils with elastic modulus within the range of 70 GPa. The elastic modulus of cellulose chain when calculated theoretically, it gave values up to 250 GPa.

However, there is no technology developed in order to separate them from microfibrils. The tensile strength of natural fibers depends upon the test length of the specimen which is of main importance with respect to reinforcing efficiency. According to the reports of Satyanarayana and Mukherjee, the tensile strength of the flax fiber depends majorly on the length of the fiber.

As compared to this, the tensile strength of pineapple fiber depends less on the length, whereas the scatter of the measured values for both are located mostly in the region of the standard deviation. The molecular fine structure of the fiber controls the properties of flax fibers, which themselves are affected by used fiber processing techniques and the conditions in which they grew. The properties of flax fiber are controlled by the molecular fine structure of the fiber which is affected by growing conditions and the fiber processing techniques used. Flax fibers have moderately high-specific strength and stiffness.

3.3 DEGRADATION OF CELLULOSE ACETATE-BASED MATERIALS

Cellulose acetate (CA) is an improved natural polymer with a huge number of properties. Cellulose acetate (CA) can be manufactured with a range of degrees of substitution (DS). The commonly used level is a DS of 2.5 because of molecular weight, melt properties and good solubility in common solvents.

Because of these properties, cellulose acetate (CA) has applications in a number of consumer products, such as films, textiles, plastics and cigarette filters that can be discarded and become litter. Say for instance the production of cellulose acetate (CA) materials globally was above 800,000 metric tons a year in 2008 and a number of items are dumped of as surface garbage or into composting facilities, it is necessary to know the fate of dumped cellulose acetate (CA) based products.

By increasing the ways of awareness for degradation could prove beneficial to know the final impact of the environment. The initial examination of biodegradability of cellulose acetate (CA) concluded that the polymer is not biodegradable, which is incorrect. This happened because the examination being conducted only with cellulose-degrading organisms like fungi. The importance of the step of deacetylation was realized after it came into existence that enzyme common in microorganisms is acetyl esterase enzyme. As per the scientific community, cellulose acetate (CA) is still a biodegradable polymer. Current work has permitted a significant growth in the knowledge of the enzymology of non-substituted and acetylated polysaccharides, thus enlightening the biodegradation mechanism.

One more commonly used mechanism for a number of polymers is Photodegradation. Cellulose acetate (CA) is the only polymer that has limited photodegradation in sunlight. Most of the consumer products consist of additives that enable enriched photodegradation. A photooxidation catalyst which allows degradation in sunlight is Titanium oxide and is used as an additive in order to increase the whiteness of cellulose acetate (CA) materials.

3.3.1 Biological Degradation

Some contradicting conclusions were presented the initial researches on the biodegradation of cellulose acetate (CA) polymer. Some of the researchers concluded that cellulose acetate (CA) with a degree of substitution (DS) of 2.5 had less use because of its degradation, whereas some researchers determined that it is impossible for the natural organisms to degrade cellulose acetate (CA) with a degree of substitution (DS) of higher than 1.5.

Lately, experiments recognized that the initial deacetylation step by chemical hydrolysis and acetylesterases was recognized as the chief mechanism for degradation. Thus, enabling the degradation of the cellulose backbone with cellulase. In hindsight, perhaps it is undeniable that the mechanisms for biodegradation should be different for cellulose versus CA polymers.

The earliest study on the biological degradation of a number of derivates of cellulose was published by Reese in 1957. With this, it was concluded, that only one acetyl per glucose was enough in order to make sure the resistance to degradation by cellulase enzymes. This study does not include the deacetylation step. In 1972 EPA Report No EPA-R2-72-046 was published investigating the biodegradability of packaging plastics. The researches included showing the polymers to fungi using ASTM Method D-1924-63 and judged the growth rates after exposure times of up to 3 weeks. The authors concluded as per the limited test conditions that cellulose acetate (CA) is not biodegradable.

The process of biodegradation was observed during the initial researches when cellulose acetate (CA) - based reverse osmosis membranes of DS 2.5 were incubated with a variety of microorganisms. Not all sources of microorganisms were successful in attacking the cellulose (CA) membranes, only some were. Cellulose acetate (CA) lost their semi-permeability with up to 50% loss of acetyl groups after the incubation for two months.

It is significant to note that some definitions of biodegradation are debatable. According to one's view is a requirement of a certain rate of degradation, such as weight loss versus time. While other says that the conversion of a substrate initiated by a microbe in a biologically active environment into methane (anaerobically), carbon dioxide (aerobically), cell wall material, and other biological products.

One novel study was the findings by forensic science researchers Northrop and Rowe in 1987, who studied the effect of the soil environment on the biodeterioration of man-made textiles. They observed that cellulose acetate fibers were decayed (but not completely) after two months in wet soil and were decayed completely after a time period of four to nine months. According to this study, no major changes were shown by other synthetic textile fibers including acrylics, polyester and nylon by the end of twelfth-month study.

As a result of these studies and investigations, it was finally confirmed that cellulose acetate (CA) is biodegradable in the natural environment as measured by a number of methods. The most convincing study about degradation was the aerobic biodegradation of radiolabeled cellulose acetate (CA) was given by Komarek, Buchanan and Gardner where they examined the release of carbon dioxide from in vitro samples with acetyl carbons labeled with ^{14}C. The research compared cellulose acetate (CA) with degrees of substitution (DS) of 1.85, 2.07 and 2.57, and found the biodegradation rates were reduced, but not inhibited, by the higher levels of acetyl.

In 1993, Gardner, Buchanan and Komarek studied two aerobic assay systems for degrading CA films: an *in vitro* enrichment cultivation technique and an activated sludge wastewater treatment system. The degradation of cellulose acetate (CA) films by enrichment culture occurred successfully within two to three weeks. This was identified by 67 percent weight loss.

Although a similar degradation was seen by industrial wastewater treatment system the degradation process took about ten weeks, which was comparatively slow. The authors emphasized on the point that the escalated systems of degradation included an increase in concentration only and no change was observed in the capabilities of the microorganisms that occur naturally. Hence, when cellulose acetate (CA) materials are kept in a natural environment, the possibilities for biodegradation remains conserved and the only change observed is the change in the speed of degradation.

Cellulose acetate (CA) can also be degraded under anaerobic conditions. The cellulose acetate (CA) of increasing degree of substitution (DS) was studied by Rivard et al. in 1992, by incubating them along with special culture for ninety-eight days and observed a considerable degradation in cellulose acetate (CA) of the degree of substitution less than 1.25. Although a degradation in samples with a degree of substitution (DS) greater than 1.7

was also observed, comparatively to a lesser extent. In the year 1993. Gu, Eberiel, Gross and McCarthy examined the biodegradability of cellulose acetate (CA) films with a degree of substitution (DS) between 1.7 and 2.5 with composting conditions that are kept under control. At a temperature of 53 ^{0}C, the materials were tested in biologically active laboratory aerobic test vessels.

As a result of incubation for seven and eighteen days respectively, they observed that the films vanished completely. In order to know the mechanism of degradation, the researchers assessed the aforesaid degraded films by proton nuclear magnetic resonance (^{1}H NMR), gel permeation chromatography (GPC) and scanning electron microscopy (SEM).

The important observations included pitting of the films by microbes and a few changes in the degree of substitution (DS) and the molecular weight while the polymer undergoes degradation. It was proposed that the random cleavage of the polymer caused low molecular weight degradation products that would diffuse away from the bulk materials and be rapidly assimilated by the microorganisms.

In this same investigation, GU et al. presented the results of keeping cellophane and a cellulose acetate (CA) film with a degree of substitution (DS) equal to 1.7 in a bioreactor under anaerobic conditions. The cellulose acetate (CA) film underwent complete degradation after a period of thirty days.

3.3.2 Enzymology of Cellulose Acetate Biodegradation

A number of important studies and investigation has been conducted into the enzymes responsible for degrading cellulose acetate (CA) in recent years. Non-derivatized cellulose can be efficiently hydrolyzed by using cellulolytic enzyme system that comprises of many cellobiohydrolases (CBH), acting from the non-reducing as well as reducing end of the cellulose chain, and a random action of endoglucanases (EG) occurs within the chain. On a commercial scale, the same enzyme system is being used for the manufacturing of the second-generation ethanol from lignocellulosic material.

Cellobiohydrolases not only have the capability to attack crystalline and amorphous regions of cellulose but are also sensitive to all kinds of cellulose substituents, comprising acetyl groups, for the reason that the active site of this enzyme class is in a tunnel formed by stable surface loops. The tunnel-shaped active site restricts the hydrolysis to occurring at the ends of the cellulose chain. This is because of the continuation of hydrolysis progressively. It can be stopped at the encounter of the first substitution. Cellobiohydrolases are supposed to be incapable of attacking cellulose derivatives. The presence of the active site inside a tunnel could account for the fact that the derivatives of cellulose would get stuck, as per their charge and their size.

Cellobiohydrolase I (CBHI) cleaves ß-1, 4-glycosidic bonds with retention of configuration, yielding the ß-anomer as a reaction product. It begins attacking from the reducing end, while cellobiohydrolase II (CBHII), and attacks the non-reducing end of the cellulose chain. CBHII inverts the configuration at the glycosidic bond, in order to obtain the formation of cellobiose first as the α-anomer.

Considering the structure of cellobiohydrolases, observing that the mixture of CBHI:CBHII:EGI:EGII = 6:2:1:1 in pure form was not able to degrade CA DS 2.5 was quite expected because the acetyl cleaving enzymes were missing in this artificial mixture. Although 43 percent of Kraft pulp that was not bleached and 78 percent of cellophane were hydrolyzed within a time period of 6 hours.

As compared to exo-acting cellobiohydrolases, it is only possible for endoglucanases to degrade disordered, amorphous regions of cellulose, cutting at internal glycosidic bonds. Although, beneficially their site is comparatively more open arranged and kept in a cleft, permitting random hydrolysis of the cellulose chain.

Correspondingly the possibility of endoglucanase-initiated hydrolysis of cellulose derivatives is positive, the extent not only being dependent on the degree of substitution, but also on the substituent distribution along the cellulose chain and the architecture of the enzyme protein. For efficient hydrolysis of CA, endoglucanase would be the most favored especially if non-substituted or partly substituted a hydro glucose unit are present.

3.4 SYNTHESIS OF BIOPLASTIC-BASED RENEWABLE CELLULOSE ACETATE

Biodegradable polymers are a good source of waste management alternative as compared to conventional plastics. Life Cycle Assessment supports this fact. In the past few years, a lot of attention has been drawn by the manufacture of biodegradable materials from renewable natural resources like plants, specifically in EU countries and the use of renewable resources has been revitalized. If managed properly, this would not only decrease their impact on the environment upon disposal but also would be applicable technically as well as economically.

It is supposed that most commonly used polymers in our day-to-day lives are, plastics, majorly in packaging applications. In 2015, the annual production of plastics based on petroleum jumped 300 million tons. In fact, every year 34 million tons of wastes comprising plastic are produced throughout the world. 93 percent of this waste is dumped off in oceans and landfills. By the foundation of bioplastics, fossil fuels are saved, carbon monoxide emissions and plastic waste are reduced. The plastics derived from the biomass like sugarcane, corn and wood are termed as bioplastics.

The following describing phase is included in the process as per the recent invention:

A plasticization phase of the cellulose acetate flake conducted in a plow mixer along with a cooling jacket to maintain the increased temperature produced by the friction between the numerous components of the formulation and due to the exothermic reaction that takes place during the plasticization phase, under control temperature i.e. below 54°C.

This thermal effectiveness ignores the action of the solvent produced by the heating of the plasticized products confined in the mixture.

In the course of plasticization phase, moreover, it is assumed initially that the hydrogen bond present between the cellulose acetate molecules is replaced by the hydrogen bond present between the molecules of the mixture of plasticizers and cellulose acetate. The energy balance of this process is slightly exothermic.

- Following are the set of steps conducted after plasticization phase: In order to generate granules, plasticization process gave birth to the extrusion phase of the plastic materials.
- the extrusion phase(s) of the granule is/are conducted at a temperature lying within the range from 150 to 200°C;
- a molding phase of the granules to attain an end-product; the molding phases of the granules for the manufacture of molded end-products are conducted at temperatures in the range of 160 to 220°C;
- a production phase or an extrusion phase as per the Compression molding process for manufacturing a product that is semi-processed for example a sheet.

- The extrusion phases of the sheets beginning from granules are conducted at temperatures lying in the range of 150 to 200°C.

Some examples of esters giving rise to the mixtures of plasticizers are- glycerol tribenzoate, acetyltriethyl citrate, tributyl citrate, triacetin monolactate, triethyl citrate, glycerol diacetate, butyl tartrate and glycerol triacetate.

The mixture of plasticizers specifically consists of acetic esters of glycerol and majorly contains glycerol triacetate (GTA), acetyl triethyl citrate (ATEC), glycerol diacetate (GDA) and low-molecular-weight (<5000) polyester based on adipic acid.

Preferably the mixture of plasticizers is a binary mixture whereas the plasticizers are an acetic ester of glycerol mixed with either an ester of tartaric acid or an ester of citric acid, most preferred a mixture glycerol triacetate (GTA) and acetyl triethyl citrate (ATEC), also a mixture of an ester of citric acid and an acetic ester of glycerol is preferred.

To obtain a binary mixture of plasticizers, it is preferable to mix the plasticizers in a weight to weight ratio of 1:1. The most preferable binary mixture is the glycerol triacetate (GTA) and acetyl triethyl citrate (ATEC) in a weight ratio of 1:1.

Specifically, the phase of plasticization includes the process of mixing cellulose acetate in a quantity that ranges from 60 percent to 90 percent by weight. Further, the cellulose acetate, having a molecular weight in the range of 40,000 to 65,000 is mixed with a mixture of plasticizers in the quantity that ranges from 10 percent to 40 percent. The most preferred mixture is the amount of plasticizer in the weight range of 20 percent to 40 percent with respect to the total weight of the mix of plasticizers and acetate. As per a standard internal method, the Molecular weight (MW) is measured by the means of Gel Permeation Chromatography (GPC).

It takes around 5 minutes to 20 minutes for plasticization phase to execute as per the standard rules. The additives mentioned below are induced preferably during the plasticization phase:

- UV-radiation stabilizing additives;
- additives to get photochromatic effects;
- additives to stabilize the process;
- IR-absorber additives (infrared rays), if possible based on phthalocyanine with a maximum IR radiation absorption at 920 nm;

- aromatization additives;
- additives for producing graphic effects and markings with laser ray application;
- Suitable compatibilizing and binding additives for using crushed or ground wood-pulp discards, preferably gained from processing waste of the wood industry, i.e. without the felling of trees and free of polluting additives of the carpentry industry.

On the basis plasticization degree, that ranges from a 10 percent (minimum value) to 40 percent (maximum value), the discarded granules possess a glass transition temperature or Tg, whose values range from 100 degree Celsius to 150 degree Celsius, being disclosed by TMA analysis. According to the international standards ASTM D-790, Flexural Elastic Modulus Values ranges from 3,000 to 1,000 MPa, according to the standard ASTM D-256, the value of notched Izod impact strength ranges from 30 J/m to 150 J/m and according to the standard ASTM D-785 the value of Rockwell surface hardness scale R ranges from 140 (minimum) to around 70.

The major source of bioplastics is biomass feedstock and agricultural resources that easily and readily fulfill with the materials that are sustainable, renewable and eco-efficient in nature. As per the reports, it has been found that global production capacity of bioplastics has grown exponentially in the past decade from 0.36 Mt in 2007 to approximately 3.45 Mt in 2020.

It is found more efficient to use cellulose as a raw material amongst all the biopolymer matrices that are being used to produce the bioplastics. It is also known that cellulose is found in all plant materials in abundance. In teak wood tree or *Tectona grandis sawdust,* cellulose is found in considerable amounts. Commonly the wood is composed of around 15 percent to 30 percent (wt. by wt.) of hemicellulose, 50 percent (wt. by wt.) of cellulose and 5 percent to 30 percent (wt. by wt.) of ash.

In order to enhance the quality of bioplastics CA is used, cellulose acetate (CA) is formed by acetylating the cellulose. Wood pulp is the main ingredient in the manufacturing of CA. It is further reacted with acetic anhydride and acetic acid in the presence of Sulphuric acid to create cellulose triacetate.

The high glass transition temperature is presented by Cellulose acetate and it very difficult to perform melt-process using CA as a raw material due to its thermal properties. Due to this reason, the need for plasticizers arise and hence they are added to CA. These substances minimize the rigidity of the polymer and allow them to melt without any degradation in their thermal properties. The biodegradable plastics where Cellulose acetate is used as a base were studied and it has been found that these plastics decompose in water or soil within a time period of few years only. Further, these plastics can be easily recycled or burned to leave no residue behind. A lot of studies were performed that evaluate the properties of CA in details like impact resistance, color ability, mechanical strength, transparency, moldability, fabricating versatility, and dielectric strength.

Another positive aspect of natural plastics is that they are formed in liquid form and hence it becomes very easy to shape them and also, they do not require a huge amount of energy. As compared to the traditional plastics, those are commonly stored in the form of granules and require a huge amount of energy in order to give them the desired shape by injection, molding or extrusion.

3.4.1 Producing Biodegradable Plastics with the help of Agricultural Wastes

The prospects of waste management treatment have broadened with the introduction of biodegradable polymers and are efficient as compared to traditional plastics. Biodegradable polymers are also supported by the Life Cycle Assessment. Landfilling is one of the worst techniques to dispose of the end-of-life while the most preferred end of life disposal technique for these materials are municipal and domestic composting.

Hence, biodegradable polymers play an important role and play a significant role in the recovery of material, utilization of renewable resources and reducing the landfills (Davis and Song, 2006). Biodegradable films are being developed and are being introduced in the commercial channel in order to overcome the drawbacks of conventional polyethylene which are very difficult to recover by mulching film post use.

Biodegradable films are generally manufactured by using bio-based materials which, after their use, can be filled in the land along with the mulched remains of plants, so as to start the process of decomposition by microorganisms. (Demetres et al., 2013)

The Bio-preferred Program initiated by the U.S.A Department of Agriculture undertook a lot of important initiatives in endorsing and promoting the procurement of bio-plastics at the federal level. In 2012, the two most commercially successful and influential biobased and biodegradable polymers were starch-based polymers and Poly-Lactic Acid (PLA), that combined account for approximately 88 percent of total biodegradable polymer consumption, separately, starch-based polymers were 41 percent and Poly-Lactic Acid (PLA) was 47 percent of total consumption (Petrova and Garner, 2014).

One other biodegradable polymer that was commercially successful is microbial Poly-Hydroxy Alkanoates (PHA) which, from the last couple of years, have been marketed as biodegradable plastics (Ying et al., 2014). According to Chen and Patel, 2012, "PHA has been marketed as environmentally friendly bio-plastics with less CO_2 emissions and sustainability as well as independence from petroleum sources". Further, it has been found that PHA exhibits numerous industrial applications as well (Viviana et al., 2014).

According to Davis and Song, 2006, "If managed properly, the development of biodegradable packaging materials from renewable natural resources (e.g. crops) has received increasing attention, particularly in EU countries" and this usage of renewable resources have been restored (Tabone et al., 2010, Cateto et al., 2008, Kiatsimkul et al., 2008).

During current years, the growth and expansion will minimize the impact on environmental after disposal (Davis and Song, 2006) and also it would be economically and technically practicable (Tanaka et al., 2008). Cellulose Acetate (CA) is used as a base for biodegradable plastics and these plastics were researched and studies it has been found that the produced plastic can be easily decomposed in water or soil within a few years only. As per Alexander (1993), these materials can be reused, and further can be incinerated without leaving any residue.

Going in further details, the proper researches and studies were performed on the significant and vital characteristics of CA, that includes impact resistance, mechanical strength, color ability, moldability, transparency, dielectric strength and fabricating versatility. (Fischer et al., 2008, Jinghua et al., 2009). If we describe the applications of CA, there are a lot many to discuss as they can be used in the production of plastic tools, ultrafiltration membranes, photographic films and fibers (Cosimo, 2013).

With the help of CA, natural plastics are manufactured in liquid form and hence, shaping the final product becomes easier and also a large amount of energy is not needed in this process. This can be compared to conventional plastics that are generally stored in form of granules and a huge amount of energy is required in order to shape the plastic by using various techniques like injection, extrusion or molding. (Xiaoyun and Shuwen, 2013). For the production of CA, a lot of researchers used acetylation of fibers found in plant cellulose, for example, by-products of cotton; straws of wheat, rice, barley and rye; poplar and cornhusk wood fiber. The process of acetylation is performed in an ionic liquid (Cao et al., 2007), or in supercritical carbon dioxide (Nishino et al., 2011), and also by iodine and acetic acid (Cheng et al., 2010) or by phosphotungstic acid (Guozhi et al., 2013).

As the costing of raw materials plays an important role on the cost of bio-based plastics and have a high impact in the production scale, using negative value cellulosic or low-cost raw materials seem quite attractive for the industrial manufacturing of CA plastics. According to Li and Renneckar, 2011, Arg-Egypt, 2011, Flax is one of the ancient crops from Egypt and flax fibers accounts for around 7 thousand tons per annum. Flax fibers constitute 92 percent cellulose (Textile Learner, 2012). Cotton linters are observed as an important cellulose raw material around the globe, for manufacturing of paper, so as to convert cellulose derivative and for revived fibers.

During the process, if cotton seeding, cotton linters are obtained as a byproduct. During the 1970s, the oil millers gain approximately 12 percent of raw cotton linter from processing cotton seeds or it can be understood as, from 1000 kg of cottonseed, around 120 kg of raw cotton linter was obtained; but in 2009, it got reduced to around 6 – 8.5 percent while cottonseed account for around 63 percent of the cotton crops (Axel, 2009).

According to a thesis submitted at Assiute University, in Egypt, the approximate quantity of the cotton crops was around 82,829 thousand tons during the years 1990 to 2008. While it has also been found that cotton linters constitute 94 percent cellulose (Textile Learner, 2012).

3.5 BIODEGRADABLE PLASTIC MATERIAL DERIVED FROM CELLULOSE ACETATE AND RELATIVE END PRODUCTS

To understand this section, first of all, let's understand what Cellulose Acetate is! Cellulose Acetate is a plastic compound, which is gained when cellulose of natural origin undergoes the process of acetylation. This is used for decades in the production of hair ornaments and fashion accessories, buttons and extensively in the manufacturing of eyeglass frames.

Production of Cellulose acetate got successful in commercial space due to several reasons like in particular, the bright colors and pleasant feel, that hugely attracted the stylists and fashion designers to translate the material in various color effects and numerous aesthetics. These properties and characteristics made Cellulose acetate a very favorable, flexible and versatile raw material of a rapidly and continuously evolving and changing the industry in the fields of altering style and taste.

Presently, however, the needs and challenges for finding a perfect raw material are very different and difficult as there are numerous factors that must be kept in mind before picking up the material. With the attention that a product is always "fashionable and up-to-date", the hunt for new modifications and products of traditional products should also consider all the hurdles that are linked with the conservation of the environment, for instance, troubles and issues related to the disposal, non-toxicity, and compatibility of the materials.

There are continuous Research and Development are going on in parallel on the conventional plastic material derived from cellulose in order to discover ways to evolve them towards the products that are more compatible and are more interesting in terms of "Aesthetics". Further, the newly evolved products shall also take care of the environment, are economically more profitable and non-toxic to the user along with higher technical performances. Also, at the same time, they should also possess upgraded aesthetical characteristics and are more essential for the category of end products where they are utilized.

For more than 80 years, cellulose acetate based plastic materials have marked their presence in the commercial market. Cellulose acetate that is derived from materials such as wood cellulose and cotton linters is mostly obtained in flakes and powder form. In required specifically, the Cellulose acetate can also be obtained in granular form.

For this, the CA powder or flakes are subjected to process called plasticization process that undergoes in the presence of appropriate plasticizers and required additives. It is apparent that both the compounds; the choice of additive and the cellulose acetate noticeably influence the manufacturing and production of compatibility, biodegradable and non-toxic end products.

Further, it is also evident that CA that compose of an acetyl substitute degree of 2.2 or less are compostable and biodegradable in the sea environment and in the soil. Also, it should also be taken in a note that higher values of acetyl substitute in CA lead to less biodegradable product. Quantity and type of plasticizer are one of the important features for making sure that the product is biodegradable in nature and plays a crucial role in deciding the biodegradation rate, in addition to possible residues deriving from the preparation reaction of cellulose acetate. Plasticizing and stabilizing additives, for example, like Diethylphthalate (DEP), conventionally used in the manufacturing of cellulose sheets and granules, discharge the substances that are hard to biodegrade for example phthalic acid, during the degradation phase.

Other more plasticizers that are used on a large scale to promote their flame-retardant properties such as Diphenylcresylphosphate or Triphenylphosphate helps in a good plasticization, but these substances can release non-biodegradable compounds like phenol, both during the processing phase and possibly during degradation process of the end products. As per the rules and regulations of the provisions of EC 1272/2008, phenol is characterized as a toxic substance and marked dangerous in the case of continued contact with the skin and mutagenic of the group.

Further, the elements that should also be taken into consideration are the ones that help in producing end products, for example, eye-glasses. They should also be taken into consideration for the problems and troubles arise through them against the conservation and safety of the environment, for example, the issues related to the compatibility and disposal of the end products. To make sure the non-toxicity of the end product as well as the raw material is also very crucial. They are in any case object to alter the processing and parameters of plastic materials that have been consolidated over the years.

It is thus extremely significant to find processes and products that not only improve the compostability and performances but also guarantee the features that allow traditional processing processes to be applied, thus upholding the flexural modulus and Tg values of the material within values very comparable to those of traditional materials. The Applicant has astonishingly found that the process for the manufacturing of plastic material based on cellulose acetate rendering to the present invention overwhelms the drawbacks defined above.

In specific, cellulose acetate driven plastic materials that are obtained on the basis of present inventions to allow the Eco-compatible or biodegradable plastic materials are obtained based on the cellulose acetate. These are used in enhancing and maintaining the compatibility of said plastic material, dimensional stability and aesthetical properties with polycarbonate, which plays an important role in the optician field, allowing the use of all types of material and lenses, also including PC, opposing the case with applications that uses traditional formulation of acetate for eye-glasses.

It is easy to relate an object of the present invention and hence it can be related to the process of manufacturing of plastic materials that are derived from cellulose acetate, as the process described is categorized by including plasticization phase of cellulose acetate, which is achieved by bringing together a mixture of plasticizers and the cellulose acetate flake or powder. This mixture of plasticizers includes of a minimum of two plasticizers selected from esters of tartaric acid, esters of citric acid and esters of glycerin while keeping the temperature below 50 degree Celsius in the plasticization phase.

3.6 A NEED FOR MORE SUSTAINABLE RAW MATERIALS FROM HEALTHCARE, CONSTRUCTION, FOOD AND COATINGS FOR CELLULOSE BASED PLASTICS

The basic definition of cellulose plastics can be that bio-plastics that are produced using the cellulose or by-products of cellulose are called Cellulose Plastics. The basic raw material required for the manufacturing of the Cellulose plastics id the softwood trees. Further, the barks of the trees are detached from core and are used as a fuel/ energy source in the process of production.

To manufacture the cellulose plastics, the cellulose fiber is separated from the tree, by heating or cooking the tree in a digester. According to a report by Transparency Market Research, lignin's and resins are obtained as a derivative in the digester. These byproducts can further be utilized as a feedstock or as a fuel in the production of other chemical products.

The outcome of the digester is a pulp that consists of alpha cellulose and hemicellulose. The pulp produced is then mixed with bleaching chemicals in order to eliminate even minute traces of lignin's and resins and to minimize the hemicellulose part from the resultant pulp. Further, this processed pulp contains a large amount of water that must be get rid of from this pulp before starting the process of mixing this pulp with high alpha cellulose content.

The next step is the manufacturing of cellulose esters from this processed pulp. This cellulose ester is used in the manufacturing of cellulose plastics. This cellulose ester is obtained by a series of steps in which this processed pulp is allowed to react with some anhydrides and acids of different concentration and at different temperatures. The concertation and temperature values solely depend on the application of the end user. The chemical composition and properties of cellulose ester highly depend on the anhydrides and acids used in the production process. Butyrate, propionate and acetate are amid the major kinds of cellulose esters. Cellulose acetate is the leading product type for cellulose esters and the drift is expected to continue during the times of forecast. Most important applications for cellulose plastics include extruded films, sheets, thermoplastics, electronics, eyeglass frames, rods, etc.

Molding materials are the most important segment for applications of cellulose plastics and the trend is probably to stay for a conceivable future. Plastic is manufactured mainly using non-renewable sources such as crude oil and its various by-products leading to the high values of the carbon footprint at the time of production of plastics.

Moreover, other concerns such as environmental and biodegradability hazards related to traditional plastics have directed to increase in a number of regulations to govern the use of plastics. The guidelines imposed on manufacturing and use of plastics has led to rush demand for bio-based plastics and that has been escalating demand for cellulose plastics.

The other foremost growth factor and a major driver in surging the demand for the cellulose plastic market have been the exponential increase in demand for electronics items, for example, screens shields and transparent dialers, etc. One of the major issues is the fact that softwood is the leading raw material required for manufacturing cellulose plastics and with an increase in the number of deforestation regulations and strict rules has become a major restraint for the market.

Low cost and easy availability of traditional plastics is also a major hurdle in the growth path for cellulose market. Further comparative cost-benefit and high efficiency of traditional plastics as compared to cellulose plastics have practically controlled the growth rates for cellulose plastics.

3.6.1 Research and Development in Cellulose Market

The cellulose plastics market holds the infinite growth potential and with continuous rising research and development in order to produce low cost and high efficiency is expected to offer massive growth prospects. According to a new study, it has been found that plant cellulose has the potential to provide a biodegradable and renewable alternative in place of polymers that are currently used in 3D printing of products.

"Cellulose is the most important component in giving wood its mechanical properties. And because it is inexpensive, bio-renewable, biodegradable and also very chemically versatile, it is used in a lot of products," said lead researcher, Sebastian Pattinson of Massachusetts Institute of Technology (MIT) in the US. Pattinson further added that "Cellulose and its derivatives are used in pharmaceuticals, medical devices as food additives, building materials, clothing, all sorts of different areas. And a lot of these kinds of products would benefit from the kind of customization that additive manufacturing- 3D printing enables".

When cellulose is exposed to heat, before becoming flowable, it starts decomposes thermally at first. Due to the intermolecular bonding, high concentration cellulose solutions start behaving like a very viscous liquid and it easily gets extruded from the solution, claimed researchers. In order to overcome this problem, the researchers opted to cellulose acetate to work upon. Cellulose acetate is a material that can be made with cellulose as a base with not much difficulty and it is already readily available and produced in large quantities around the globe.

The major advantage of using cellulose acetate is that it reduces the number of hydrogen bonds with the help of the acetate group. Further Cellulose acetate can be extruded through a nozzle and can be dissolved in acetone. It is known that acetone evaporates quickly, leaving behind the solidified cellulose acetate in place.

Annual renewable biomass and agricultural feedstocks highly affect the rate of production of these materials. Further, there is an immense potential for the biopolymers derived from natural sources to overtake the commercial space that is currently enjoyed by the products and items exclusively derived from petroleum feedstock. As every coin has two sides, biodegradable polyesters like PHA and PLA also have some drawbacks as compared to oil-based plastics, for example, low glass transition temperature (Tg), low fracture toughness, difficult processing along with limited applications, low fracture toughness and high moisture absorption.

The most significant biopolymer derived from cellulose, Cellulose diacetate (CDA), from a commercial point of view, is a thermoplastic material developed with the help of esterification of cellulose. Cellulose acetate has been one of the commercially successful bio-based and biodegradable bioplastic, so that new applications of CDA can be predicted for the specific tasks like packaging and automotive but also pharmaceutical and medical applications, in the manufacturing of bio-related and other bio composite fields.

Different type's raw materials, for example, recycled paper, cotton, sugar cane, and cellulose are utilized in the manufacturing of cellulose esters in different forms like Powder. One of the shortcomings of CDA is that it has high glass temperature (Tg), due to which it has limited scope for processability as compared to the conventional plastics used in making of commodities. Along with it, CDA has very less solubility in most of the general solvents and is not recommended as a raw material for melting processes as it starts to decompose before the stage of melting starts.

As it is evident from the above mentioned case that it is not easy to work with CDA, when in a natural state, hence, some additives like plasticizers are used to reduce the melting temperature and to reduce the Tg. It is very necessary to add these for CDA processing. The function of plasticizers is to reduce the intermolecular forces among the polymer chains, that results in a more flexible and softened matrix of the polymer.

Conventionally, for the process of plasticization of CDA, the main ingredients used were phthalates, citrate, triacetin, glycerol derivatives, phosphates, etc. One of the main and most common plasticizers for CDA was Phthalate esters, extensively used in industries have been found to be harmful to the environment and also as a health threat and thus with further research, it has been found that there are severe concerns attached to phthalate esters when used in long term.

So as to advance the processing speed of CDA, some research and development work was carried out and few new plasticizers were explored for example polyethylene glycol, poly (caprolactone triol), dibutyl phthalate and propylene glycol.

Other plasticizers were obtained by combining glycol, TA and maleic anhydride in different proportions. With the advanced research and development, the processing of CDA based materials has increased considerably but the price paid for that was the poor mechanical properties of the final output. By studying and examining the literature in detail, CDA processing was done along with added TA in a particular study, as it is considered to be a plasticizer that is considered environment-friendly and sustainability proof due to its fast biodegradability and low toxicity.

Further, tri-acetan is a plasticizer which is highly active for CDA because of its high solubility in CDA. The minimum corresponding value of Gibbs energy obtained by mixing of CDA/TA blend is about 40 wt. percent TA, due to the fall in the values of entropy of the system. This process is immediately followed by the formation of solvates between the functional groups of Triacetin and CDA. The increase in the system's entropy happens due to the disturbance created by the interchain interaction during the Solvation in CDA. Beyond this amount, both the main mechanical properties (tensile strength, modulus, and elongation at break) of materials became very low and Gibbs energy of the mixing increases. In order to improve mechanical properties, cellulose diacetate was blended with Poly (N-vinyl pyrrolidone – co-vinyl acetate), Poly (vinyl acetate), Poly (vinyl pyrrolidone) and Poly (caprolactone monoacrylate).

A lot of trials were undertaken to spread the cellulose diacetate chain with the help of reaction with octadecenyl succinic anhydrides and also with another bio polymer for example starch and Polylactic acid. Most of the cellulose diacetate related to research is used as a popular solvent to dissolve the materials and then go on with processing.

This process created a porous structure and also abridged the material's toughness. Some of the researchers a different approach called "melt processing approach" to enhance the performance of cellulosic plastics. Further, a lot of studies were researched and tried in order to come up with biocomposite derived from cellulose diacetate with the help of natural fillers and fibers, while some other researchers considered using clay which is organically modified so as to develop "green" and "eco-friendly" nanocomposites.

The compatibility and factors affecting the durability/ suitability of biopolymers in particular and cellulose fibers for thermoplastics, in general, were studied thoroughly as they possess a lot of high mechanical properties, dimensional stability, low density, and ecological advantages. According to Glasser et al, "the preparation of composites based on continuous rayon and lyocell fibers reinforced cellulose acetate butyrate (CAB) and other commercially-available thermoplastic cellulose esters by using solution impregnation."

To alter the interfacial adhesion between the cellulose ester matrix and cellulose fiber the fibers were also modified on the surface by acetylation, but there was no noteworthy rise in the strength of the composites was notices.

On the basis of these results, the researchers lead to the final conclusion that interfacial stress transfer cannot be seen as a shortcoming in this system, further, these outcomes and results can be easily understood by anybody that possess enough knowledge in the field of mechanics to perform the fiber reinforcement of the composites. Further, until recent times, the features and characteristics of composites developed with the help of cellulose esters reinforced with short lyocell fibers are not taken into consideration.

REFERENCES

1. Ach, A. (1993). Biodegradable Plastics Based on Cellulose Acetate. *Journal of Macromolecular Science, Part A*, [online] 30(9-10), pp.733-740. Available at: https://www.tandfonline.com/doi/abs/10.1080/10601329308021259?journalCode=lmsa20 [Accessed 28 Mar. 2019].
2. Ganster, J. and Fink, H. (2013). Cellulose and Cellulose Acetate. *Bio-Based Plastics*, [online] pp.35-62. Available at: https://onlinelibrary.wiley.com/doi/10.1002/9781118676646.ch3 [Accessed 28 Mar. 2019].
3. Kalia, S., Dufresne, A., Cherian, B., Kaith, B., Avérous, L., Njuguna, J. and Nassiopoulos, E. (2011). Cellulose-Based Bio- and Nanocomposites: A Review. *International Journal of Polymer Science*, [online] 2011, pp.1-35. Available at: https://www.hindawi.com/journals/ijps/2011/837875/ [Accessed 28 Mar. 2019].
4. Mostafa, N., Farag, A., Abo-dief, H. and Tayeb, A. (2018). Production of biodegradable plastic from agricultural wastes. *Arabian Journal of Chemistry*, [online] 11(4), pp.546-553. Available at: https://www.sciencedirect.com/science/article/pii/S1878535215001100 [Accessed 28 Mar. 2019].
5. Patents.google.com. (n.d.). *Biodegradable plastic material based on cellulose acetate and relative end-products - Google Patents*. [online] Available at: https://patents.google.com/patent/EP2599827A1 [Accessed 28 Mar. 2019].
6. Phuong, V., Verstiche, S., Cinelli, P., Anguillesi, I., Coltelli, M. and Lazzeri, A. (2014). Cellulose Acetate Blends - Effect of Plasticizers on Properties and Biodegradability. *Journal of Renewable Materials*, [online] 2(1), pp.35-41. Available at: https://www.researchgate.net/publication/264742062_Cellulose_Acetate_Blends_-_Effect_of_Plasticizers_on_Properties_and_Biodegradability [Accessed 28 Mar. 2019].
7. Plastemart.com. (2017). *Cellulose-based plastics address need for more sustainable raw materials from food, healthcare, coatings and construction*. [online] Available at: http://www.plastemart.com/plastic-technical-articles/cellulose-based-plastics-address-need-for-more-sustainable-raw-materials-from-food-healthcare-coatings-and-construction/2356 [Accessed 28 Mar. 2019].
8. Puls, J., Wilson, S. and Hölter, D. (2010). Degradation of Cellulose

Acetate-Based Materials: A Review. *Journal of Polymers and the Environment*, [online] 19(1), pp.152-165. Available at: https://link.springer.com/article/10.1007/s10924-010-0258-0 [Accessed 28 Mar. 2019].

9. Rohmawati, B., Atikah Nata Sya'idah, F., Rhismayanti, R., Alighiri, D. and Tirza Eden, W. (2018). Synthesis of Bioplastic-based Renewable Cellulose Acetate from Teak Wood (Tectona grandis) Biowaste Using Glycerol-Chitosan Plasticizer. *Oriental Journal of Chemistry*, [online] 34(4), pp.1810-1816. Available at: http://www.orientjchem.org/vol34no4/synthesis-of-bioplastic-based-renewable-cellulose-acetate-from-teak-wood-tectona-grandis-biowaste-using-glycerol-chitosan-plasticizer/ [Accessed 28 Mar. 2019].
10. Thanh Phuong, V., Lazzeri, A. and Cinelli, P. (2012). *"Green" Biocomposites Based on Cellulose Diacetate and Natural Resources - Cellulose Microfibers and Lignin*. [ebook] Available at: http://www.escm.eu.org/eccm15/data/assets/977.pdf [Accessed 28 Mar. 2019].

Chapter 4

CHITIN AND CHITOSAN

CONTENTS

In the current times, it has become really important to utilize the wastes that are generated in the biological forms to manufacture products that can be easily recyclable and reusable and those that have a property of being bio-degradable. The environment comes under serious threat due to the chemically manufactured synthetic products which also increases the importance of such biodegradable material. Chitin and Chitosan are such materials that are very hard in nature and have very wide range of uses in several industries. This chapter informs the readers about these compounds, their manufacturing processes, their uses and applications and the importance they hold in the present industrial setup.

4.1 INTRODUCTION

Chitin is known as one of the most abundant polymers of amino polysaccharide present in nature. It is the basic building material which provides the potential to the exoskeletons of insects, crustaceans, and cell walls of fungi. Through chemical as well as enzymatic degradation, chitin can easily be converted to its well-known derivative named chitosan.

The efficient natural sources of chitin are crab and shrimp shells, which are considered to be a rich byproduct of the food industry that gives the huge quantity of this biopolymer to be utilized in the biochemical applications. In chitin synthesizing organisms, the synthesis and deprivation of chitin need a tough enzymatic control in order to sustain homeostasis.

Chitin synthase which is a pivotal enzyme present in the pathway of chitin synthesis utilizes UDP-N-acetylglucosamine (UDPGlcNAc) and produce the polymer of chitin. On the other hand, chitinase enzymes degrade the chitin.

Bacteria are known as the chief mediator of the degradation of chitin present in nature. Chitin and chitosan having ideal biochemical properties like biodegradability, biocompatibility, ability to form films, non-toxicity, etc. have gained various biomedical applications as well.

The field of nanotechnology has increasingly used chitin and chitosan-based products in its recent advancements. Chitin and chitosan have widely been used so as to fabricate the scaffolds of the polymer. In addition to that, the application of chitosan in order to yield designed nanocarriers and to enable the techniques of microencapsulation is under growing investigation for the delivery of biologics, drugs and vaccines.

Every single application is tending to need ideally designed chitosan-based nanoparticles with definite dimensions as well as cargo-release features. The capability of reproducibly produce chitosan nanoparticles that can condense the protein cargos with great loading efficiencies is still a major challenge.

Chitosan can successfully be used in various solutions such as Nano- or microparticles and hydrogels, and an infinite array of derivatives with modified biochemical properties, with distinct degrees of deacetylation, can be prepared.

4.2 WHAT IS CHITIN AND CHITOSAN?

Chitin (β-(1–4)-poly-N-acetyl-D-glucosamine) is extensively distributed in nature currently. After cellulose, it is the second most ample polysaccharides available. Chitin occurs in nature in the form of ordered macro fibrils. It is the key component in the exoskeletons of shrimps, crabs, crustaceans and the cell walls of fungi.

For numerous biomedical applications, chitin is converted to chitosan which is known as the deacylated derivative. Both chitin and chitosan are biodegradable, non-toxic and biocompatible biopolymers. Also, they are hydrating and antimicrobial agents.

Based upon the chitin source, it occurs in the form of two allomorphs, viz. the α and β forms, that can further be categorized by infrared as well as solid-state NMR (Nuclear Magnetic Resonance) spectroscopy, together with the X-ray diffraction. γ-Chitin is a third form of allomorph after α and β.

These allomorphs basically differ in the alignment of microfibrils. Biosynthesis of chitin is usually catalyzed by chitin synthase, which is known as the highly conserved enzyme present in nature. This enzyme is present in every organism that synthesizes chitin. Chitin synthase remains destined to the increasing chain of the polymer by various steps of polymerization that consecutively adds GlcNAc units to the non-reducing end of the growing chain.

The first product obtained is linear polymers. These linear polymers then assemble into different microfibrils of different length and diameter. After the biosynthesis of polymer, the final fibrils are transferred to the extracellular space. Glycosyl hydrolases (GH) are an extensive class of enzymes that are used in the degradation of polysaccharides. Based upon the resemblances in their amino acid sequences, they are broadly classified into 130 families. Chitinases are mainly classified into GH-18 and GH-19 families. They can be easily found in various organisms such as fungi, bacteria, insects, animals, and plants.

Chitin and chitosan are widely used in biomedical applications, mainly in gene and drug delivery, tissue engineering, stem cell technology, etc. These kinds of biopolymers can easily be processed into numerous products like membranes, hydrogels, beads, nanofibers, nanoparticles or microparticles, sponges and scaffolds.

Amongst all of the biomedical applications, tissue engineering is the field where chitin and chitosan are used effectively so as to fabricate the polymeric scaffolds for the aim of tissue regeneration after its repair.

The most significant features of these biopolymers that have made them most preferred candidates for the fabrication of polymeric tissue scaffolds are biodegradability, high porosity, and structural integrity, predictable rate of degradation, biocompatibility and non-toxicity to cells. Loading nanoparticles based on chitin with several drugs like 5-fluorouracil and lamivudine, are just two key examples of effective application of chitin and chitosan in the field of drug delivery.

Recently, quantum dots or semiconductor nanocrystals have been coated with chitosan for the use of bioimaging in the field of cancer diagnosis. The improvement of immune responses against various pathogenic microbes by applying chitosan derivatives as vaccine adjuvants have also been reported by the experts.

4.3 CHITIN

Chitin can be described as a large and structural polysaccharide, which is made up of modified glucose chains. Chitin can be found in the exoskeleton of insects, plants and animals, and in the cell walls of fungi, and in hard structures of fish and invertebrates. Chitin comes at the second position in context to abundance.

In our biosphere, more than 1 billion chitins are produced every year by various organisms. This versatile molecule can itself make solid structures as in wings of insects and can also combine with various other components such as calcium carbonate to make tough substances such as the clam shell.

No vertebrate animal can digest chitin, like cellulose. Animals that take the diet of insects usually have protozoa and symbiotic bacteria, which can easily break the fibrous chitin into glucose molecules. However, as chitin is one of the biodegradable molecules that is dissolved over time, it is utilized in various industrial applications such as binders and surgical threads for glues and dyes.

4.3.1 Structure of Chitin

Chitin is mainly a long chain polymer (a linear homopolysaccharide) comprising repeated units of N-acetyl-glucosamine. N-acetyl-glucosamine is known to be a monosaccharide derivative of glucose. These units further form covalent β-1,4 linkages. Chitin having the chemical formula $(C_8H_{13}O_5N)_n$ is considered as an intricate carbohydrate, whose structure is similar to that of cellulose, with one hydroxyl group on each of the monomer replaced by an acetyl amine group.

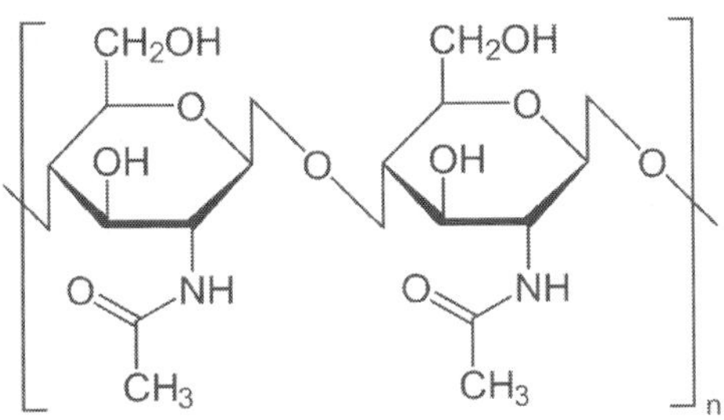

Figure 4.1: Molecular Structure of Chitin.

Source: https://upload.wikimedia.org/wikipedia/commons/thumb/3/3b/Chitin_Haworth.svg/2000px-Chitin_Haworth.svg.png

Chitin comprises modified monosaccharides of glucose. Glucose basically exists as a ring of oxygen and carbon molecules. Bonds present between the glucose molecules are called glycosidic bonds. The oxygen that usually forms hydroxyl groups linked to the carbon ring can form a bond with another carbon instead of hydrogen. In this manner, monosaccharides can be bonded together in the long chain. Chitin is formed by a sequence of glycosidic bonds between distinct substituted molecules of glucose.

Due to the substitution that takes place on the molecule of glucose, chitin is different from cellulose. The glucose molecule present in chitin has an amyl group instead of a hydroxyl group and it comprises carbon as well as nitrogen. Nitrogen is an electrically positive molecule whereas the oxygen molecule which is double bonded to the group is negative.

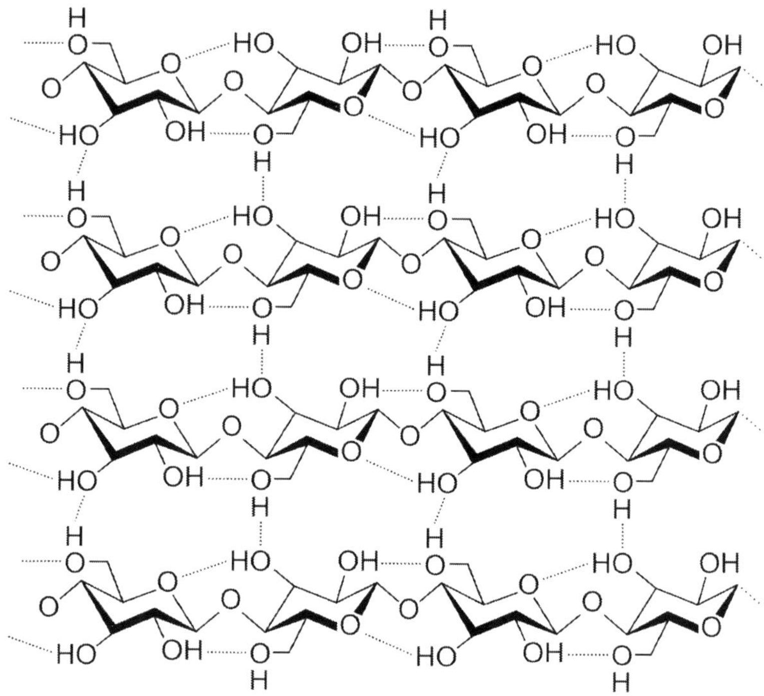

Figure 4.2: A strand of cellulose.

Source: https://upload.wikimedia.org/wikipedia/commons/thumb/a/a4/Cellulose_strand.svg/1131px-Cellulose_strand.svg.png

This eventually produces a dipole in the molecule which intensifies the hydrogen bonds that are formed between these molecules and the molecules around them. When united in a matrix with several compounds and molecules of chitin, the ultimate structure can be quite hard because of all the weak interactions between neighbor molecules.

4.3.2 Function of Chitin

Like keratin and cellulose, chitin is a structural polymer. It is basically made up of small monomers or monosaccharides. When chitin is secreted in and out of the cell in an organized manner, the fibers then form weak bonds between each other. This further provides the potential to the overall structure. Chitin, as well as cellulose, are made up of the glucose monomers, whereas keratin is a kind of fibrous protein.

Numerous structural polymers arose during the initial phase of the evolution of life, as they are found in few groups only. Keratin is exclusive to animals, cellulose to plants and chitin to the fungi, mollusks and arthropods. Chitin, as well as cellulose, grew very early in the history of life, whereas keratin arose in few animals long after fungi and plants.

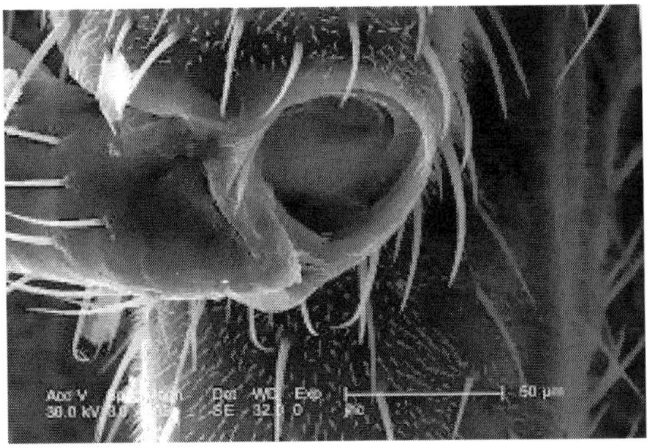

Figure 4.3: Microscopic view of the exoskeleton of chitin-containing organisms.

Source: https://encrypted-tbn0.gstatic.com/images?q=tbn:ANd9GcSdxKPLm_0jetjTGsvG8LdnGZKMHAsikxQ5XqMAp2x79FVx2GDA

Chitin is also known as the naturally occurring polymer which forms a polymer. It plays an effective role in protection in various lower eukaryotes. Chemically, chitin is a long chain polysaccharide which is unbranched and made up of residues of N-acetylglucosamine bonded by β-1,4 covalent bonds.

By considering the role played by the chitin molecule in various biological structures (insect peritrophic matrix, fungal cell walls, insect and crustacean cuticles, cyst wall of protozoa, eggshells from nematodes), its metabolism that is biosynthesis and degradation is important for distinct morphogenetic events.

Chitin is not present in plants and vertebrates; however, it participates in the interactions of host-parasites and further represents a target specific to the parasite for the chemotherapeutic attack. Chitin also plays a major role in various host immune responses.

Figure 4.4: Microscopic view of the outer skeleton of an insect.

Source: https://upload.wikimedia.org/wikipedia/commons/thumb/6/64/Antlion_11793.jpg/1024px-Antlion_11793.jpg

Due to its extreme presence in nature and its ideal properties, biotechnological uses of the derivatives of chitin are currently a growing area in pharmaceutical, biomedicine, Agro-biosciences and food technology.

The skeleton present on the outer part of the body appears rigid and hard because of the presence of chitin which is known for its elastic properties, However, chitin is the leading constituent. Numerous other compounds like calcium carbonate and proteins play a significant role in the exoskeleton formation. The key role of the chitin-containing exoskeleton is to keep soft tissue safe from any kind of injury.

Most substantially, it averts these soft tissues from being dry. In brief, it behaves as a watertight barrier against the dehydration, which is very critical for their existence. The hard chitin-containing exoskeleton of the arthropods behaves as a mechanism of defense against predation as predators apply a compressive force on the exoskeleton to hurt their victim. The fungal cell wall which protects the micro-organisms from the outer environment is also comprised of chitin.

4.3.3 Uses of Chitin

As a Fertilizer

One of the significant advantages of chitin is its utilization in preparing fertilizers. Fertilizers having chitin are non-toxic, organic and they are observed to enhance the productivity of crops. Chitin present in fertilizers also helps in increasing the enzyme activities and soil organisms that affect the health of soil positively and thus increases the yield of crops.

As a Food Additive

Chitin has really long history of application as an additive of food. Generally, it is obtained from shellfish that has shrimp, and crabs. Few of the times, cell walls of the eumycete, which is a kind of fungi, are utilized as a source for chitin extraction. Microcrystalline chitin (MCC) being a food additive can be very helpful in order to improve the flavor as well as taste.

As an Emulsifying Agent

Using food chitin can also aid in developing stable emulsions of food. Essentially, it acts as a good emulsifying agent which further helps to avert the emulsion break when exposed to various other fluids. For example, the toppings of whipped dessert generally have chitin that gives stability as well as uniformity to the product.

Medicinal Uses

The naturally occurring and fiber-forming polymer that is chitin can also aid to decrease the levels of cholesterol as found out by several animal studies. The molecules of chitin tend to swab the cholesterol as well as fat present in the digestive system. Hence, chitin present in the diet may help in decreasing the absorption efficiency of the cholesterol.

As a Surgical Thread

Chitin is also utilized for making flexible and strong threads that can be further used surgically. A few dissolvable stitches applied to close the wound are made up of chitin. These stitches get decomposing during the process of wound healing. Different reports by the experts recommend that stitches that are made up of chitin may also help in facilitating the wound healing.

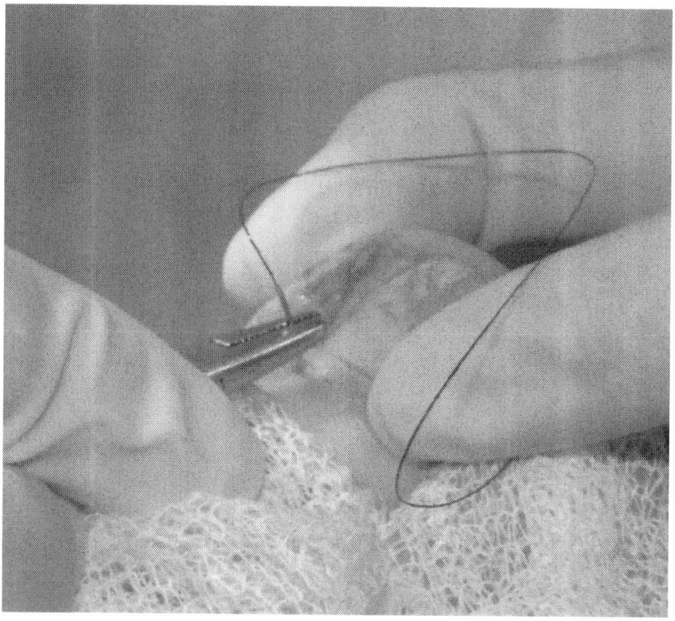

Figure 4.5: Chitin as a surgical thread.

Source: https://upload.wikimedia.org/wikipedia/commons/b/b7/WedgeResection_All_16.JPG

4.3.4 Examples of Chitin

Chitin in Arthropods

Arthropods may be perceived as a species that varies widely in its existence. These are the kinds of animals that do not possess a vertebral column and have a body layout that is divided into different parts. They possess an outer skeleton that is very hard in nature and is comprised of chitin and other different proteins. This kind of layout of the body proves to be highly advantageous for the Arthropods and is useful for them across various ecosystems.

These organisms can be found in all parts of the ecosystem irrespective of the altitude and climatic variations. They may exist in the extreme depths of the oceans and seas and may even be found in the highest altitudes that can support existence. The size they exhibit may also vary significantly as they have been found to exist in the form of mites which can be found at the pores of the hair and in the form of huge crabs and insects, whose length may be given in meters.

Figure 4.6: Chitin can be found in the wings of insects.

Source: https://upload.wikimedia.org/wikipedia/commons/5/59/Vespula_germanica_Richard_Bartz.jpg

The main composition of the skeleton of all these arthropods comprises of chitin which is found in unison with the structural proteins. By combining with different kinds of proteins, chitin takes the shape of a very flexible substance and thus, is instrumental in shaping up the wings of various insects.

It is this ability of chitin to transform and adapt to various materials that has enabled the arthropods to develop themselves and take various different shapes in the form of different species accounting to more than a million.

Chitin in Fungi

Chitin may be utilized to develop the cell wall in fungi. Chitin can be found in the extracellular form in combination with different proteins and other elemental structures, in a form similar to that of cellulose in plants.

This shapes up to develop a very firm wall in between the cells which enable the organisms to maintain the shape they have acquired. In a manner, similar to the one in plant cells, the cells may reserve the water in themselves in order to develop a water pressure that may be distributed evenly against the cell wall.

This water pressure helps to strengthen each of the cells by a greater magnitude and is referred to as the Turgor pressure. The fungi have the ability to come out of several layers of leaf litter during the time they are growing, and the grown layers can weigh a lot up to a large number of pounds. This develops as a matter of strength that chitin possesses in the form of the structural fiber.

Chitin in Mollusks

Chitin may be viewed as a wide range of various other ways in the mollusks. The usage of chitin can be found both in the mollusks that are at lower levels and those that are in the form of better-derived cephalopods.

In the creatures referred to as the mollusks, which may include organisms such as snails, the chitin may represent a part of an organ that resembles a tongue in spiked form, known as the radulae. The creatures that possess these features utilize the radulae in order to get rid of the different kinds of food forms such as algae, from surfaces it develops on, which are hard in nature.

The more derived forms, which are the cephalopods, make use of chitin in a way that they develop a beak so that it can break through the protective walls of the items it considers as its prey as the walls are generally quite hard in nature. This hard nature of their preys can be attributed to them being arthropods themselves, which explains the material of the composition of their shell, which is chitin.

4.3.5 Biological Importance of Chitin

The plants possess Cellulose, which is particular to their composition. In a similar way, the animals possess a specific material, known as Keratin. Carrying the exclusiveness forward, the insects possess Chitin, which can also be found in the cell walls of some fungi. The composition of Chitin boasts of a modifies glucose, which makes it a large polysaccharide.

It showcases its importance in the following ways:

- It provides a hard, external cover for the insects to protect them from the attacks
- An animal with a vertebral column is generally not able to digest the cellulose form, which is Chitin. Hence, the insects are secured from their predators in this regard
- Some arthropods make use of such material to prey on their food,

such as in case of snails.

- The compound, Chitin, due to its cellulose and hard form finds itself very useful in the industries. It is used to manufacture surgical threads and the binding component in the dyes and adhesives.
- As stated above, Chitin is used to maintain the water pressure against the cell walls in the fungi, in order to strengthen the cell from within.

Figure 4.7: Various insects possess Chitin in their skins.

Source: http://footage.framepool.com/shotimg/qf/206565895-water-beetle-water-surface-europe-continent.jpg

4.3.6 Importance of Chitin in the Marine Environment

Chitin proves to be an important and significant form of deposits of carbon and nitrogen, that is made available to the various organisms in the marine ecosystem due to the high abundance of the component in the marine ecosystem.

The procedure in which the degradation of Chitin takes place serves as an important stage in the distribution of the nutrients in the marine ecosystem and there is an important role that chitinolytic bacteria has in the cyclic process.

These kinds of bacteria originate from both types of ecosystem, whether marine or freshwater and give rise to chitinases that help in the degradation of chitin. This polysaccharide is insoluble in nature and with the use of chitinases, they are broken down into biological components that later can be useful in the ecosystem.

4.3.7 Health Benefits and Uses of Chitin

Chitin can be represented by chemical formula $(C_8H_{13}O_5N)_n$, depicting a chain of molecules of N-acetylglucosamine. This compound is widely known as the main component in the external structure of the arthropod skeleton.

The term arthropod is generally used to refer to organisms such as shrimps, lobsters, crabs and various kinds of shellfish, mainly the ones that can be classified into the categories of insects and crustaceans. The main role that chitin plays can be associated in similarity to the one played by Keratin in the vertebral animals, which is to help in the development of hair, nails, and feathers.

In 1929, a chemist from Switzerland, Dr. Albert Hofmann, showcased the building blocks of chitin. The building blocks of chitin are same as that of cellulose, which resembles a polysaccharide that is made up of linkages of small blocks of glucose, which may also be categorized as monomers.

The hydroxyl group that is found in the monomers is separately substituted by the acetyl amine compound ($COCH_3NH$), in order to form chitin. Hence, the high strength of this compound in comparison to cellulose is justified, as the replacement tends to give rise to the hydrogen bonding in the monomers separately, transforming the strength to a great extent. The pure forms of chitin are very flexible in nature, and also maintain the high amounts of strength and toughness they develop. Though, generally the toughness of chitin is increased even further to support some causes, by addition of a foreign compound.

An instance of this can be found in the insects, the outer skeleton of which is composed of chitin mixed with sclerotic, which is a compound with a crisscross structure of proteins. Another example can be found in the shells of crustaceans, which are mainly made up of the combination of chitin and calcium carbonate.

There are several uses of chitin, even in the applications that help in generating huge commercial gains such as those of the food industry and the products related to pharmaceutical production. Chitin is generally utilized as a thickener in foods and as a stabilizing agent. It is also used as an edible substance for covering of the food.

The compound, chitin can be used as a dietary fiber that is used as health supplements and serves the main purpose. There exists a transformed form of chitin in the form of chitosan, that is found in better availability in the ecosystem and have a large number of uses as health supplements. This form of chitin, that is chitosan, is manufactured on an industrial scale by drenching crustacean shells in a highly concentrated alkaline solution of sodium hydroxide. This is the process that helps in separating the acetyl group ($COCH_3$) in chitin.

4.4 CHITOSAN

Chitin and its deacetylated subsidiary chitosan are polymers in the natural form made from haphazardly dispersed β-(1-4) - connected D-glucosamine (deacetylated unit) and N-acetyl-D-glucosamine (acetylated unit). Biopolymers such as chitin and chitosan display different properties that open a wide range of uses in different segments, particularly, in biomedical science.

The most recent developments in the research related to biomedical science are essential developing patterns that hold an incredible guarantee in items related to the management of healing the wounds. Chitin and chitosan are considered as valuable biocompatible compounds to be utilized in therapeutic equipment that is used for treatment, improve or supplant any tissue, organ, or working of the body.

An assemblage of ongoing examinations proposes that chitosan and its subsidiaries are promising compounds for supporting products in tissue building applications. The research and current developments are primarily centered around the contemporary research on chitin and chitosan towards their applications in various biomedical fields to be specific tissue building, counterfeit kidney, skin, bone, ligament, liver, nerve, muscle, healing the wounds, treatment of burns and a few other helpful purposes.

Chitin is a cellulose-like biopolymer found predominantly in the outer skeleton of marine creatures, for example, shrimp, crabs, or lobsters. Chitin can likewise be found in mushrooms and yeasts. Chitosan is a synthetically processed type of chitin. "Squid pens," squander shell that forms as a result of the processing of squid, are a sustainable and cheap way of extracting chitosan.

4.4.1 Structure of Chitosan

Chitosan is a straight polysaccharide made out of arbitrarily appropriated β-(1→4) - connected D-glucosamine (deacetylated unit) and N-acetyl-D-glucosamine (acetylated unit). It is made by treating the chitin shells of shrimp and different crustaceans that comprise of chemicals that are alkaline in nature, similar to sodium hydroxide.

4.4.2 Chemical Properties of Chitosan

The chemical properties that chitosan exhibits are as per the following:

- Linear polyamine,
- Various amino groups that are reactive in nature,
- Accessible groups of hydroxyl compounds that are reactive in nature,
- Chelates numerous transitional metal particles.

4.4.3 Biological Properties of Chitosan

Following are the natural properties of chitosan:

- Biocompatible,
- Natural polymer,
- Biodegradable to ordinary body constituents,
- Safe and non-lethal, (the study in chitinase is critical in this regard),
- Binds to mammalian and microbial cells forcefully,
- Regenerative impact on connective gum tissue,
- Accelerates the development of osteoblast in charge of the bone arrangement,
- Hemostatic,
- Fungistatic,
- Spermicidal,
- Antitumor,
- Anticholesteremic,
- Accelerates bone arrangement,
- Central sensory system depressant,
- Immunoadjuvant.

4.4.4 Industrial Applications of Chitosan

Because of its physical and synthetic properties, chitosan is being utilized in an immense range of broadly distinctive items and applications, extending from pharmaceutical and restorative items to water treatment and plant assurance. In various applications, distinctive properties of chitosan are needed. These properties may change with the level of acetylation and weight of the molecules.

Cosmetics

Normally natural acids are utilized as great solvents for uses in the cosmetic industry. A characteristic amino poly-saccharide, chitosan can be incorporated in the class of hydrocolloids. Nonetheless, being a lot different from the other different hydrocolloids which are polyanions, chitosan is the sole cationic gum, being natural, that ends up being thicker on being treated with acid. It encourages its cooperation with general integuments (skin spreads) and hair. Chitin and chitosan are fungicidal and fungistatic in natural form. Chitosan is good with groups of naturally dynamic segments consolidated in the formation of cosmetic items.

Chitosan or chitosan-alginate composites in the scope of 1-10 μ, just as microcapsules including different hydrophobic substances find a lot of uses in cosmetic items. It might be noticed that substances engrossing the destructive UV radiation or diverse colors can be effectively covalently connected to chitosan amino groups.

Structures dependent on chitosan and different hydrocolloids containing enemies of oxidants, hostile to unfavorably susceptible, and calming substances of vegetable birthplace, new sorts of depilatory and means for twisting and doing the hair are being worked out by the people serving in Sonat Co., USA. Chitin, chitosan and their subordinates offer use in three zones of beauty care products: hair care, healthy skin, and oral care.

Chitosan and hair are integral to one another attributable to convey inverse electrical charges: chitosan positive and hair negative. A reasonable arrangement that contains chitosan shapes an unmistakable, versatile film on hair, in this way expanding its non-abrasiveness, smoothness, and mechanical quality.

The material can likewise shape a gel when added to mixtures of alcohol and water. Chitosan can be utilized in shampoos, washes, lasting wave agents, hair colorants, styling creams, hair splashes, and hair tonics.

A few subordinates of chitosan and chitin have potential uses in hair care. They incorporate glyceryl chitosan, an adduct of an oligomer of hydrolyzed chitosan, n-hydroxypropyl chitosan, quaternary hydroxypropyl-substituted chitosan, polyoxyalkylene chitosan, chitosan oligosaccharides, chitin sulfate, and carboxymethyl chitin. A few subordinates of chitosan can shape froth and make emulsifying activity and chitin powder can be utilized in cleanser without any other help.

Chitosan and its subsidiaries have two favorable circumstances that make it a great contender for healthy skin: one being their positive electrical charge, and then another that the sub-atomic weights of most chitosan items are high to the point that they can't enter the skin.

In this manner, for example, chitosan can work as a cream for skin. In view of its lower costs, it may contend with hyaluronic acid in this application. Both chitosan and chitin are as of now found in creams, pack material, salves, nail polish, nail finishes, establishment, eye shadow, lipstick, purifying materials, and shower specialists. Chitosan acylated with a natural diacid anhydride, and fine particles of chitin or chitosan are utilized for healthy skin.

Both chitin and chitosan can be utilized in toothpaste, mouthwashes and chewing gum. They refresh the breath and keep the growth of plaque and tooth decay away. Salts of chitosan, added to toothpaste, cover the horrendous taste of silicon oxide and combine powders with the goal that they keep up their granular shapes. Chitin can likewise be connected as a dental filler material and both chitin and chitosan ingest candida like thicans, a parasite that sticks to teeth, making them a contender to clean false teeth.

Water Engineering

Because of its polycationic nature, chitosan can be utilized as a flocculating agent. It can likewise go about as chelating agent and overwhelming metals trapper. Weltroswki and few other researchers utilized chitosan N-benzyl sulphonate subordinates as sorbents for evacuation of metal particles in acidic medium.

In 1999, Bhavani and Dutta reported the expulsion of hues from dye house effluents utilizing chitosan as an adsorbent. Extensive measures of world creation of chitin and chitosan and subsidiaries are utilized in wastewater treatment. Chitosan molecules agglomerate, to a great extent, anionic wastes in the quest for the formation of precipitates and floe, thus it goes about as a woolly to reuse of food processing waste. Chitosan can contend viably with engineered tars in the capture of substantial metals from processing water.

Chitin has been utilized to clean plutonium containing wastewater, and water containing methyl-mercury acetate, huge contamination of wastewater from acetaldehyde generation. Utilization of chitosan/chitin blend was found to expel arsenic from polluted drinking water.

Chitosan has additionally been discovered as being successful in the expulsion of oil and oil-based commodities from wastewater. The deacidifying capacity of chitin is used in espresso industry and to clear up the drinks, for example, wine, beer, and natural product juices.

Recovered chitin, chitosan, and different chitinous layers could be generally utilized for such procedures as osmosis, reverse osmosis, filtration for micro compounds, desalination, dialysis, and hemodialysis. Beds of chipped chitosan can likewise be utilized for cleaning of consumable water.

Paper Industry

Biodegradable chitin and chitosan can reinforce reused paper and give rise to a more environmentally friendly way of packaging products and different items. Chitosan is, as of now, associated with the assembling of paper considering that chitosan atoms look incredibly like those of cellulose, which is the principal constituent of plant walls.

It additionally spares synthetically added substances and increases the yield. In conclusion, the paper created with chitosan has a smoother surface and is progressively impervious to moisture. In addition, chitosan is an extraordinary incentive in the creation of bathroom tissue paper and for wrapping paper and cardboard.

Hydroxymethyl chitin and other water solvent subordinates are helpful end subsidiaries in papermaking. It may well be utilized as a biodegradable packaging material as wrapping material for food and different items.

Material Industry

Subordinates of chitin have been developed and used to grant antistatic and soil repellent attributes to the materials. In material industry, chitin can be utilized in the provisions for printing and finishing, while the chitosan can remove colors from color handling effluents. Besides, chitin and chitosan both have made wonderful contribution to medicinal related material sutures, strings, and filaments.

Food Processing

Utilization of chitosan in the food industry is outstanding mostly because it isn't dangerous for warm-blooded creatures. Microcrystalline chitin (MCC) indicates great emulsifying properties, unrivaled thickening, and gelling agent that can be used in the stabilization of food. It is likewise utilized as a dietary fiber in the baked items.

The utilization of MCC tackled a portion of the issues, for example, flavor, shading, and time span of usability, presented by different sources of fiber. It could be of exceptional significance for assembling protein-invigorated bread, even without such raw materials as emulsifiers and shortenings.

Chitin and chitosan go about as strong help for the capture of entire microbial, creature, or plant cell immobilization. Chitin has been utilized in immobilization of chemicals. It tends to be utilized as a non-absorbable transporter for very focused food raw materials, for example., (Food colors and supplements).

In India, the use of chitin in poultry records a 0.5 percent decline in the food utilization proportion and raises body weight by 12 percent in examination with birds being given a chitin free eating regime. Also, thorough investigations in the US have demonstrated that chicks benefited from a diet plan containing dried whey and chitin, used whey more proficiently and put on more weight than those that were given a diet that was comparative but chitin free.

Preliminaries additionally demonstrated that little measures of chitin added to the weight control plans of chicks and calves empowered the creatures to process milk lactose through expanded development of explicit intestinal microscopic organisms. These microorganisms obstruct the development of different kinds of living beings and create the catalyst required for lactose absorption. This property might be of huge significance since some specific groups of human and numerous other organisms have intolerance towards lactose and its products.

There is no conclusive research on the digestion of chitin and chitosan in the human body, and thus, the utilization of these polymers in food processing businesses should still be investigated further.

Farming

Chitin treated seeds (wheat) were found to have development quickening and development upgrading impacts. Chitinous augmentations to the processing mixtures or soil brought about a noteworthy decrease in root-knot worm pervasions and concealment of parasitic pathogens.

Photo

In shading photography, chitosan has been utilized as a raw material agent for the acid colors in gelatin and furthermore goes about as a guide to improve dispersion, an essential technique in creating photos.

Chromatographic Separations

Chitin and chitosan find wide assortments of utilization in chromatographic separations. The nearness of free – NH_2 groups, essential – OH groups and optional – OH groups in chitosan makes it as a helpful chromatographic agent.

Utilization of chitosan in thin layer chromatography for detachment of nucleic acids has likewise been accounted for. Rhee e, have utilized chitin and chitosan as sorbent material to strong stage extraction of phenol and chlorophenols by utilizing High-Performance Liquid Chromatography (HPLC).

4.5 CHEMICAL STRUCTURE AND PROPERTIES

Chitin is poly-(1→4)- β-connected N-acetyl-D-glucosamine. The individual sugar units in its structure are pivoted 180° as for one another, and each pair frames the disaccharide N,N′-diacetylchitobiose $[(GlcNAc)_2$.

The individual polymer chains can be depicted as helices, in which each sugar unit is rearranged concerning its neighbors. Such a structure prompts high soundness as the unbending strips are associated by 03-H→05 and 06-H→07 hydrogen bonds.

Chitin likewise has three diverse crystalline allomorphs: the α-, β-and γ-frames. These vary in the arrangement of the microfibrils. The commonest type of chitin is α-chitin. Its unit cell is made out of two N,N′-diacetylchitobiose units resulting in two chains in an antiparallel course of action.

In this manner, nearby polymer chains keep running in inverse ways, held together by 06-H→06 hydrogen bonds, and the chains are held in sheets by 07→H-N hydrogen bonds. This gives a measurable blend of - CH_2OH introductions, identical to a large portion of the oxygen atoms on every buildup, having the capacity to shape between and intramolecular hydrogen bonds.

This ends up in two distinct kinds of amide group; all are engaged with shaping the interchain C=O→H-N bonds, while half of the amide groups also fill in as acceptors for 06-H→O=C intramolecular hydrogen bonds. Arrangement of these intermolecular hydrogen bonds prompts a significant stable structure. The polymer anchors, in the long run, offer ascent to microfibrils without anyone else's input group on the off chance that they are permitted to take shape.

β-chitin is a less basic type of chitin, in which the unit cell is a N,N′-diacetylchitobiose unit, giving a polymer settled as an unbending strip, equivalent to α-chitin, by 03→05 intramolecular H-bonds. The chains in this structure are held together in sheets by C=O→H-N H-bonds between the amide groups and by the - CH_2OH side chains, which prompts arrangement of inter-sheet H-bonds to the carbonyl oxygens on the nearby chains (06-H→07).

This gives a structure of parallel poly-N-acetyl glucosamine chains with no inter-sheet H-bonds. The parallel course of action of polymer chains in β-chitin takes into account more adaptability than the antiparallel formation found in α-chitin, however, the resultant polymer still has tremendous quality. γ-Chitin is the third allomorph, having blended parallel and antiparallel formations.

It has been found to occur in mushrooms. Chitin is constantly discovered cross-connected to other basic parts except for the β-chitin found in diatoms. Chitin is found covalently attached to glucans in contagious cell walls, either straightforwardly, as in Candida albicans or through peptide spans.

In addition, in insects and different invertebrates, chitin is constantly connected with distinct proteins, with both covalent and noncovalent bonding. This affiliation implies it creates the observed arranged structures.

There are additionally differing degrees of mineralization, for example, calcification, and sclerotization, including collaborations with phenolic and lipid particles. In life forms like parasites and spineless creatures, there have been accounted for changing degrees of deacetylation, giving a continuum of structure between chitin (completely acetylated) and chitosan (completely deacetylated).

Albeit either acids or soluble bases can be utilized to deacetylate chitin, the way that glycosidic bonds are increasingly powerless to acid which would demolish the chain, the antacid deacetylation process is utilized every now and again.

N-deacetylation of chitin can either be performed by heterogeneous or homogeneous response mixtures. The qualification among chitin and chitosan with various degrees of deacetylation isn't accurate. Setting a couple of special cases aside, characteristic chitin happens related to other auxiliary polymers like proteins or glucans, which frequently contribute over half of the mass in chitin-containing tissue.

Chitin can be N-deacetylated to such a degree, that it ends up dissolvable in dilute acidic and formic acids. In chitin, the acetylated units win, and the degree acetylation is naturally 0.90, while chitosan is a completely or mostly N-deacetylated subsidiary with an ordinary level of deacetylation of more than 0.65.

Numerous explanatory apparatuses have been utilized to decide this level of deacetylation including IR spectroscopy, pyrolysis gas chromatography, gel saturation chromatography and UV-vis spectrophotometry, ^{1}H NMR spectroscopy, ^{13}C strong state NMR, warm examination, different titration plans, acid hydrolysis, HPLC, division spectrometry techniques and all the more as of late close infrared spectroscopy.

4.6 CHITIN BIOSYNTHESIS

Chitin biosynthesis and cell handling are an exceedingly mind-boggling, multi-faceted and between the associated arrangement of occasions which begins intracellularly and finishes in the incorporation of chitin in outside supra-macromolecular structures, for example, arthropod fingernail skin and growths cell walls.

The entire procedure envelops a few individual stages:

- Consecutive biotransformation of sugars (primarily trehalose or glucose). This progression incorporates biochemical responses like phosphorylation, amination and furthermore development of the enzymic substrate, UDP-N-acetylglucosamine.
- Chitin synthase (CS) mixtures the chains. The CS unit is a chemical which is a piece of a protein/starch group including intently topologically stuffed particles. Such a course of action guarantees mixes of beginning chitin polymers into a crystalline fibril.
- The introduction of chitin particles which have long chains.

- Polymer translocation over the plasma film.
- Crystallization and arrangement of microfibrils by between chain hydrogen holding.
- Relationship with arthropod cuticular proteins or with different sugars in parasitic cell walls.

Chitosan is cytoplasmic macrovesicles with CS movement. These microvesicles have been recognized through electron microscopy considers utilizing contagious frameworks. The wealth of chitosan at the hyphal tip suggests their vital job in CS dealing with pre-decided areas.

Chitosan starts from organelles, for example, endoplasmic reticulum and Golgi. Chitosan vesicles contain zymogenic CS groups. After in the long run combination of the chitosan with the plasma film, the CS units become enacted through proteolytic responses. CS inclusion into plasma layers includes the intercession of focusing on and acknowledgment proteins. The accurate apparatus of such combination alongside the proteins included stays uncertain.

Chitosan-like structures likewise have been accounted for in sans cell framework got from insects; in any case, it has not been cleared up whether they are engaged with *in vivo* chitin arrangement. Additionally, chitosan-like vesicles have not yet been depicted in ideal arthropod epidermal cells.

REFERENCES:

1. Biology Dictionary. (n.d.). *Chitin - Definition, Function, Structure and Examples | Biology Dictionary*. [online] Available at: https://biologydictionary.net/chitin/ [Accessed 26 Mar. 2019].
2. Dima, J., Sequeiros, C. and Zaritzky, N. (2017). Chitosan from Marine Crustaceans: Production, Characterization and Applications. *Biological Activities and Application of Marine Polysaccharides*. [online] Available at: https://www.intechopen.com/books/biological-activities-and-application-of-marine-polysaccharides/chitosan-from-marine-crustaceans-production-characterization-and-applications [Accessed 26 Mar. 2019].
3. Guibourg, M. (2018). *Applications and uses of Chitosan and Chitin*. [online] Cuantec.com. Available at: https://www.cuantec.com/blog-and-news/blog/108-applications-and-uses-of-chitosan-and-chitin-2 [Accessed 26 Mar. 2019].
4. Hamblin, M. and Elieh-Ali-Komi, D. (2016). *Chitin and Chitosan: Production and Application of Versatile Biomedical Nanomaterials*. [online] PubMed Central (PMC). Available at: https://www.ncbi.nlm.nih.gov/pmc/articles/PMC5094803/ [Accessed 26 Mar. 2019].
5. Islam, S., Bhuiyan, M. and Islam, M. (2016). Chitin and Chitosan: Structure, Properties and Applications in Biomedical Engineering. *Journal of Polymers and the Environment*, [online] 25(3), pp.854-866. Available at: https://link.springer.com/article/10.1007/s10924-016-0865-5 [Accessed 26 Mar. 2019].
6. Kumar Dutta, P., Tripathi, V. and Dutta, J. (2004). *Chitin and chitosan: Chemistry, properties and applications*. [online] Niscair.res.in. Available at: http://www.niscair.res.in/sciencecommunication/researchjournals/rejour/Jsir/Fulltextsearch/2004/January%202004/JSIR-vol%2063-January%202004-pp%2020-31.htm [Accessed 26 Mar. 2019].
7. Martínez, J., Falomir, M. and Gozalbo, D. (2014). *Chitin: A Structural Biopolysaccharide with Multiple Applications*. [online] Els.net. Available at: http://www.els.net/WileyCDA/ElsArticle/refId-a0000694.html [Accessed 26 Mar. 2019].
8. Nicks, J. (n.d.). *Chitin: Structure, Function, and Uses*. [online] BiologyWise. Available at: https://biologywise.com/chitin-structure-function-uses [Accessed 26 Mar. 2019].

9. Puvvada, Y., Sukhavasi, S. and Vankayalapati, S. (2012). *Extraction of chitin from chitosan from exoskeleton of shrimp for application in the pharmaceutical industry*. [ebook] Available at: http://www.icpjonline.com/documents/Vol1Issue9/06.pdf [Accessed 26 Mar. 2019].
10. Shahidi, F. (2007). Chitin and chitosan from marine by-products. *Maximising the Value of Marine By-Products*, [online] pp.340-373. Available at: https://www.sciencedirect.com/science/article/pii/B9781845690137500168 [Accessed 26 Mar. 2019].
11. Souza, C., Almeida, B., Colwell, R. and Rivera, I. (2011). The Importance of Chitin in the Marine Environment. *Marine Biotechnology*, [online] 13(5), pp.823-830. Available at: https://www.ncbi.nlm.nih.gov/pubmed/21607543 [Accessed 26 Mar. 2019].

Chapter 5

ARBOFORM AND THE PROCESSING

CONTENTS

ARBOFORM can be defined as a thermoplastic granule, which is made up of lignin. Further, lignin is a compound, which makes around 30% of natural fibers and wood plant. They can be easily processed by the method of extrusion, calendaring, pressing or deep drawing and injection molding. The products of ARBOFORM are biodegradable or compostable and do not have any toxic ingredients such as MDF boards or laminated wood. The chapter deals with the brief description of lignin and its modifications, along with its composites. At the middle of this chapter, the processing and mechanical properties of ARBOFORM have been described. Later, the innovative aspects such as environmental, social and economic are described.

5.1 INTRODUCTION

Concentrated or determined attempts by the research industry and institutions have been ineffectual to create high benefits to the secondary product of the industries related to pulp and paper, which are the natural polymer lignin.

Mills of chemical pulp collect approximately more than fifty times, one hundred and six tons of it in mass, in each year all across the world. Nevertheless, a bunch of researchers and developers, created a community of compounds is known as ARBOFORM. The polymer lignin being the primary element of this new division of engineering materials that are completely based on the raw materials that are renewable in nature.

ARBOFORM is appropriate to equipment parts in industry and properties of the ARBOFORM enable it to be substituted to a limited extent or degree for plastics and processed wood. Nevertheless, it shows wood-like qualities, engineering technologies of standard polymers can process the material as a thermoplastic material. The polymer ARBOFORM can be utilized for several industrial products, by means of injection molding, extrusion and compression molding.

Materials go under process at a lower temperature. After the processing, materials are used for synthetic thermoplastic. In addition, this process does not require compounding, as a result, this saves a considerable amount of energy and cycle time. The resulting elements show a little reduction or contraction than those materials that are made from the synthetic plastics, reveal outstanding acoustic properties and allow direct recycling. Sustained or no-stop research and development of the material upgraded the material.

Advancement of the materials also enhances the basic properties of the material. As anticipated from the engineering of plastics, these materials are included of high rigidness and impact strength, surface smoothness, other several operations such as flame retardancy, thermal and electrical conductivity, different colors and the absence of the processing agents.

Specifically, the process of pyrolysis can create materials inspired by advanced bio. The process of pyrolysis helps to maintain the shape at a smaller dimension. Particular samples offer an overview of different implementation or utilization of mass consumer and industrial products. These industrial products created earlier presently under detailed evaluation.

Compounds of fiber-reinforced plastics which are commonly made of glass, carbon and aramid fiber strengthened or amalgamated with epoxy and unsaturated polyesters. These compounds of fiber-reinforced plastic have played the main role through the course of the last several decades in different implementation because of their high mechanical strength and modulus.

The use of the materials, as well as dumping of these materials, more specifically glass-fiber-reinforced plastic (GFRP), have been becoming serious issues because of their non-biodegradability due to increasing environmental consciousness as well as the demands of statutory authorities associated to their recycling.

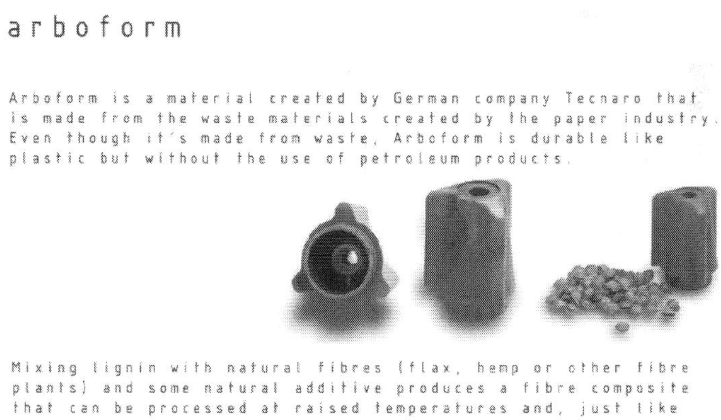

Figure 5.1: ARBOFORM.

Source: http://comafat.blogspot.com/2011/03/blog-post.html

Currently, fully biodegradable compounds can be strengthened with natural fibers that are completely natural compounds. Moreover, they have attracted more interest to solve these types of issues. Natural fiber like jute, flax, ramie, and sisal (1)-(10) are commonly applied as support function in green composites. On the other hand, biodegradable resins manufactured of byproducts from cellulose, starch, lactic acid, etc. are used as the matrix material.

Properties like biodegradable of both the elements make them easily cast out of or compounded at the ends of their useful lives. Benefits of natural fibers over other strengthening materials such as glass fiber are their low cost, recovery of enhanced energy, low density, biodegradability and acceptable strength qualities.

However, these natural compounds are not strong as the traditional glass-fiber-reinforced plastic (GFRP), the modest mechanical properties are appropriate for the implementation or utilization in non-durable consumer products and material for packing, etc. In addition, the hollow tubular design of the natural fibers decreases the bulk density of these polymers.

Consequently, compounds constructed from them are predicted to be light in weight. However, the prime disadvantage of natural fibers is that the property of hydrophilic. Hydrophilic properties of these natural fibers decrease the compatibility with the hydrophobic polymeric matrix through the course of merged fabrication.

As an outcome, the deficient fiber-matrix hold stimulates low mechanical qualities. Another disadvantage is the low processing temperature that must be used because of the other option of the thermal deprivation of fiber, which might influence the properties of the compound.

It is essential to advance the mechanical and other qualities in the natural fibers before the fabrication of compounds. Moreover, the fact that the high cost of the biodegradable resins, currently might constrained their implementation or utilization for natural compounds should not be disregarded. On the other hand, these costs are predicted to reduce in the coming years.

5.2 LIGNIN MODIFICATION AND LIGNIN-BASED SURFACTANT

The complex organic polymer known as Lignin consists of both hydrophilic operational groups as well as lipophilic carbon backbone, signifying a special attraction to aqueous and organic phases. Consequently, lignin is predicted to be a good applicant as a surfactant (compounds that lowers the surface tension between the two liquids).

Although, the challenge is constrained lignin solubility in both aqueous phase of neutral or low pH solvents as well as the most organic solvent and monomers, which slows down its practical implementation.

Presently, numerous scientific evaluations have been directed on the behavior of lignin. These evaluations have been conducted to control the amphiphile nature of lignin by modification and as a result, advanced lignin shows good potential in the area of scattering and emulsification as the bio-based surfactants.

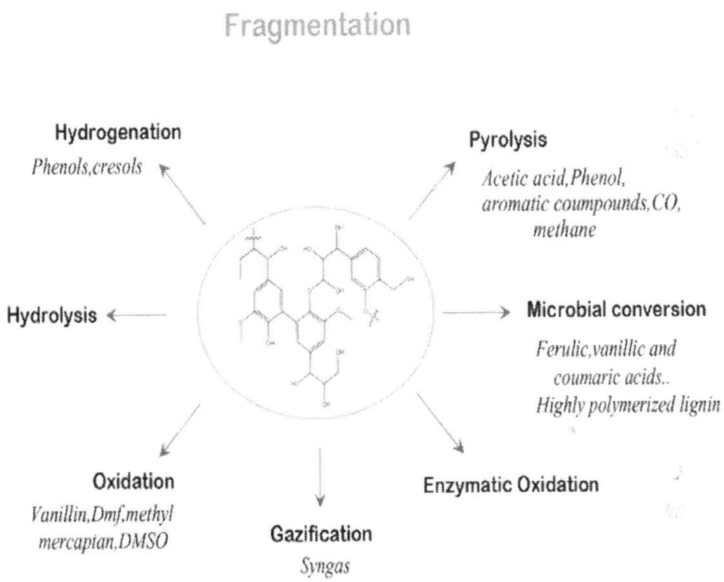

Figure 5.2: Modification of lignin.

Source:https://www.sciencedirect.com/science/article/pii/S0079670013001421

5.2.1 Lignin Modification by Chemicals

Even though, the complete pattern of structure for various types of organic polymer lignin are not same, most of the lignin consist of some common functional groups, which includes hydroxyl, methoxyl, carbonyl, ether and the family of carboxylic acids. The range of operational groups merged with its carbon backbone may cause solubility issues, while also offering the opportunity of different types of modifications or advanced approaches, like esterification, acylation, halogenation and sulfonation.

In order to tune the physical as well as chemical properties, various advanced approaches have been processed to implement the functional groups onto lignin. The family of hydroxyls in craft lignin is produced by the process of fragmentation of macromolecules of lignin with the help of bonding cleavage of the phenyl propane units in Kraft cooking digester by sodium hydroxide (NaOH) and sodium bi-sulfate (Na_2S).

There are both, the groups of aliphatic as well as the groups of phenolic hydroxyls, that are present in lignin. These groups can be operational by reactions with various chemicals. Alkylation and alkoxylation modification can be done without any difficulties on these hydroxyl areas with the help of nucleophilic aromatic substitution reaction. The resulting lignin is a by-product that can be used as pre-polymer through the course of preparation of co-polymer.

Nitration of lignin can be accomplished by reacting with nitric acid and the as-obtained lignin, and this product is commonly used as a putty of matrix of polyurethane. With amine and formaldehyde, lignin amination can be processed with the help of the Mannich reaction and the advanced lignin shows great potential as a cationic surfactant as well as strengthens putty of compounds of polymer.

The other crucial and efficient advancement method is the process of esterification. The process of esterification can transform the hydroxyl groups into esters. Acids, acid anhydrides and acidic compounds have been verified to be good agents for esterification.

With the purpose of making novel macro polyols, propylene oxides are used to esterify lignin along with the alkaline catalyst. In the area of resins which are lignin based, lignin phenylation is carried out to advance its behavior of reactivity chemically by acquainting with more hydroxide groups on the structure of lignin.

Characteristically, the conforming qualities of the advanced lignin can be adjusted by changing grafting chemical, conditions of reaction and extent of esterification. For instance, a complete esterification of hydroxyl groups in kraft lignin by butyric anhydride makes lignin very soluble in styrene. On the other hand, raw kraft lignin is not soluble at all.

5.2.2 Lignin-based Surfactant

In the distinctive molecules of lignin, there are both hydrophilic phenolic or aliphatic hydroxyl groups as well as a lipophilic carbon backbone, recommending a special attraction to polar and nonpolar phases as a surfactant. Never the less, a positive advanced stage is sometimes important to control its amphiphilicity and solubility additionally.

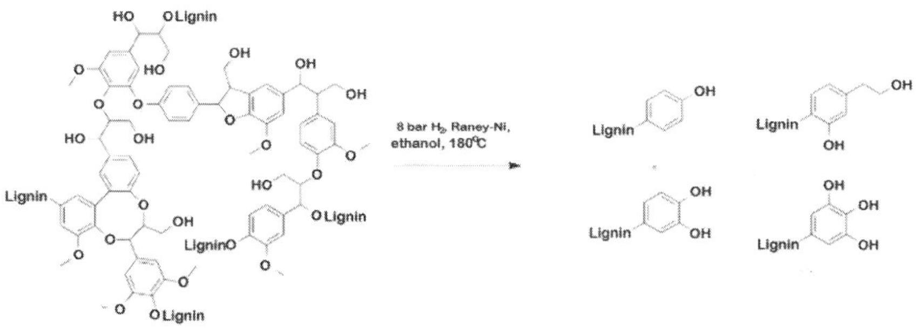

Figure 5.3: Lignin-based polymeric surfactant.

Source:https://www.sciencedirect.com/science/article/pii/S0032386117301593

Recently, copious research has been directed to control the amphiphilicity of lignin so that the outcome by-products can be functional in the area of dispersion and emulsification as bio-based surfactants.

These obtained sustainable surfactants would not only spread the implementation or utilization of lignin, but will also show better communion with the environment as compared to the traditional chemical surfactants.

The consequential lignin by-product effectively decreases the serious combination of concentration in gypsum by two to four orders of scale. Few researchers came together and arranged a water-soluble lignin-based surfactant with the help of the alkylation with 1-bromododecane in alkali pyridine/isopropanol/water system.

They found that the surface tension of water or air was reduced from almost seventy-three percent to approximately twenty-eight percent mN·m-1 when the obtained lignin by-products was applied. Further, Gupta et al. grafted polyacrylamide and poly (acrylic acid) onto kraft lignin with the help of reversible addition-fragmentation chain-transfer (RAFT) polymerization. The Rafted lignin was applied to make mixtures of water and hexane where the low grafting density in the advanced organic polymer, lignin, was conjectured to favor separation into hexane side of the oil-water interface.

5.3 LIGNIN-POLYMER COMPOSITES

5.3.1 Polymer Blending

Blending of polymers is one of the most usually functional methods to formulate various compounds of polymer with specific qualities. With stiffness in the structure and high polarity, lignin has been verified to show great potential as an organic element in the field of blends of polymer.

The mixing of organic polymer lignin and polystyrene (PS) were constructed by melting mixing in an internal batch mixer, and they demonstrated the physical properties as well as chemical properties of the resultant blends with various lignin components. For example, blending lignin with poly (ethylene terephthalate) also known as PET with the help of a single-screw extruder.

The effect of lignin on the process of crystallization of Poly-ethylene-terephthalate and the thermal properties of the lignin as well as Poly-Ethylene-Terephthalate composites were studied. Another example is of, blending the 25 lignin with the polyethylene oxide, also known as PEO synthesized by John et al.

Where lignin and Poly Ethylene Oxide were miscible because of the bonds of hydrogen molecules between these elements. In point of fact, the miscibility (the property of two substances to mix in all proportions, forming a homogeneous solution) of lignin with several other polymers is ineffective, which seems to be a big task which delays the further advancement of lignin or blends of polymer.

The affinity of lignin with several other polymers and shows that in order to acquire the desired blends of the polymers, researchers required to make cautious selections on both technical lignin type as well as the structure of polymer. The bad miscibility of lignin and polymer in the blends is resulted from the inefficient interfacial binding between these two elements.

To find out the solution for this problem or task, advancement in chemical properties is carried out to repair the structural affinity of lignin, which is supplementary employed in the thermoplastic amalgams.

Advanced lignin with maleic anhydride first and organized the blends of commercial polypropylene (PP) with the sub-obtained lignin and found that its mechanical properties had a lesser amount of fall as related with the blends of Polypropylene and the raw organic polymer lignin.

In addition, poly (methyl methacrylate) imbedded with lignin was blended with polyethylene which is low-density polyethylene (LDPE) by RRN Sailaja and the resultant compounds displayed the enhanced thermal properties and mechanical properties. Films of Nano-compounds blended with poly (lactide) also known as Polylacticacid (PLA) and improved lignin could be found by casting in chloroform. The addition of lignin was found to increase the elongation at break and toughness.

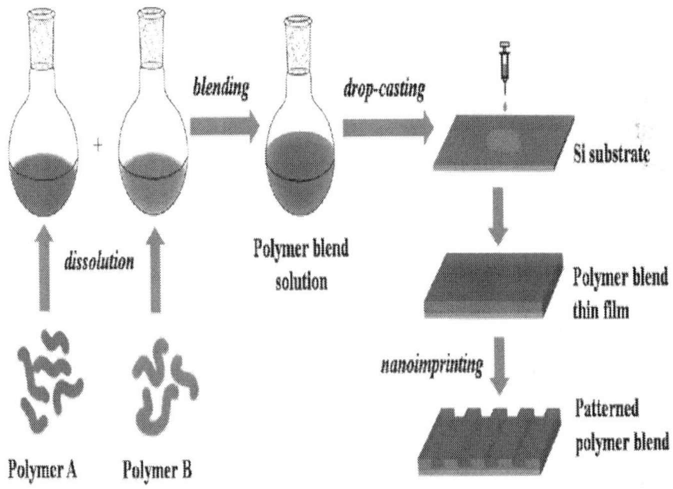

Figure 5.4: Formation of polymer blend of two mononuclear metallopolymers.

Source:https://www.researchgate.net/figure/A-diagram-showing-the-formation-of-the-polymer-blend-of-two-mononuclear-metallopolymers_fig2_325222340

Alongside, the materials made of thermoplastic, lignin is also established to show potential as strengthen agent in the rubbery polymers. Alkyl esters were used to enhance lignin in the work and the sub-obtained lignin was blended with poly (ε-caprolactone) or Polycaprolactone (PCL).

It was found that when the proportion of lignin and Poly-Caprolactone was one is to one, the mechanical properties of the compounds, which includes strength and elongation at break, were enhanced considerably. The styrene butadiene rubber (SBR) when reinforced with lignin, where the organic polymer lignin was first advanced or enhanced into lignin-layered double hydroxide compounds with the help of situ method. In this investigation, lignin was demonstrated to be well spread in the compounds by various classification methods or tactics. In addition, the most important mechanical properties of these compounds were enhanced.

Lignin or polymer blends can also be utilized in other field of operations, which includes UV protection, antioxidant as well as carbon precursors. Researchers reported that the antioxidant movement of lignin or PP blends would reduce at first and then rise with the increasing of lignin content.

In addition, in the case of lignin Polylactic acid (PLA) blends, the addition of lignin into Polylacticacid would produce the antioxidant skills; on the other hand, pure Polylacticacid showed no such type of qualities or activities. The phenol groups in lignin can serve as the radical scavengers and therefore, lignin owns the potential as the agent for UV protection, which has been demonstrated by blending with Polycaprolactone PCL through high-energy ball milling directed by Rachele et al.

Beyond, the blend of Polycaprolactone PCL and lignin Poly methyl methacrylate (PMMA grafted lignin) was found to be biocompatible, which could be utilized as one of the biomedical physical quantity.

Lignin is also a good predecessor of activated carbon in the energy storage area. Researcher constructed the battery made up of lithium ion electrodes with the help of the carbonization of lignin or Polyethylene glycol PEO blend, which showed good ability of conductivity as well as electrical properties.

5.3.2 Chemical Grafting

With the help of lignin molecules as monomers by grafting polymerization on lignin is one more approach to synthesize the compounds of lignin-polymer.

In the meantime, lignin has a relatively high Glass Transition Temperature (Tg) because of the stiff aromatic structure, the adjacent polymers of lignin with some soft elastomeric polymers may cause thermoplastic properties of which the mechanical property is offered by the organic polymer- lignin and the rubbery features are from the soft synthetic polymer.

The polyurethane, which is lignin-based, is one of the most significant examples. The fusion of lignin-based polyurethane can be comprehended with the help of grafting of isocyanate groups onto hydroxyls group field of lignin and constructing the urethane linkages by mixing lignin, diisocyanate and another diol monomer or polymerizing lignin with isocyanate-based prepolymers.

Studies states that, the thermal properties of lignin- Polyethylene glycol PEG-based polyurethane and found their Glass Transition Temperature (Tg) could be adjusted by varying the content of lignin. Saito et al. discovered that the tensile strength of the as-prepared polyurethane would enhance as they increased the lignin content up to seventy-five percent.

In the atom transfer radical polymerization (ATRP) and ring opening polymerization processes, lignin molecules are used as both one copolymer component as well as macro-initiators.

The copolymer of kraft lignin and N-isopropylacrylamide was manufactured with the help of a two-step process. In that process, lignin was first enhanced with 2-bromoisobutyryl bromide and then the ATRP was carried out with the improved lignin as initiator.

The quality of solubility of the outcome copolymer material could be managed by the ratio of lignin and N-isopropylacrylamide. Beyond, researcher demonstrated that ATRP could be applied to make Nano-particles of lignin-polymer compounds with an average diameter of five nm which also showed enhanced mechanical properties, including glass transition temperature (Tg) and modulus.

Lignin-poly (lactide) or polylacticacid (PLA) compounds can be manufactured with the help of ring opening polymerization. On the initial purpose of increasing the compatibility of lignin and the other polymer as well.

The mechanism or process of strengthening and mechanical enhancement of the lignin-PLA polylacticacidcompounds were additionally studied by Yang et al. and they stated that the crazing, the stereo complex and the rubber phase were the most important aspects which gives lead to in the process of betterment of these polymers.

Moreover, reversible addition-fragmentation chain transfer (RAFT) polymerization was also used to graft polyacrylamide and poly (acrylic acid) onto kraft lignin. The attained lignin was applied to construct the emulsions of water and hexane as surfactant. Researchers stated that, the process of copolymerization of lignin and vinyl acetate with $(NH_4)_2Fe\ (SO_4)_2{\cdot}6H_2O$ as a catalyst in the aqueous phase.

5.4 GREEN COMPOSITES: VERSATILE MATERIAL FOR FUTURE

Humans have been utilizing the materials since the beginning of the recorded history. At first, only natural materials such as stone, clay, wood etc. were used. Materials have a significant role in the life of modern human beings and have been making an important participation to the techno-economic development of the modern world.

These materials have been categorized into three main categories:

- Metals,
- Ceramics, and
- Polymers

Compounds, which are based on metal, ceramic, and polymer matrices, have also been industrialized. Polymers (synthetic as well as biopolymers) and polymer compounds have arisen as an important material. The two most important elements of a polymeric compound are:

- A polymeric matrix, and
- A reinforcing material.

A compound provides a better property than those of its fundamental materials, because of the synergistic properties of its constituent materials. In a composite, one of the constituent materials is in continuous phase, and is termed as the matrix, and the other is discontinuous or dispersed phase and is termed as the reinforcement or the reinforcing material.

Polymeric compounds are consisting of, mica or any other particle strengthened polymer compounds; short aramid fibers or any other short fiber strengthened polymer compounds; continuous carbon fiber strengthened with epoxy or any other thermoset or thermoplastic polymer compounds; nanoparticles or nanofibers strengthened with polymer compounds.

Capabilities of a polymeric compound are ruled by properties and quantity of the polymeric matrix; properties (such as aspect ratio, chemical nature, purity, distribution, orientation and geometry) and quantity of the support; and the interfacial bond between the two. Consequently, the properties of a compound are strongly impacted by the properties of their fundamental materials, their quantities and allocation, and the fiber or matrix interfacial bond.

There are more than a few benefits of these polymeric compounds. Never the less, the most important limitation is the issue of suitable removal or recycling after the end of life and reuse of both the elements quite tough. In addition, synthetic polymers (such as polypropylene, polyethylene etc.) production needs a significant utilization of resources which are based on oils, and which are especially non-renewable and as an outcome of reduce in oil reserves and increase in cost of manufacture of fossil fuels. Scientists and researchers are now discovering another or substitute sources that are renewable, environment friendly and are sustainable as well.

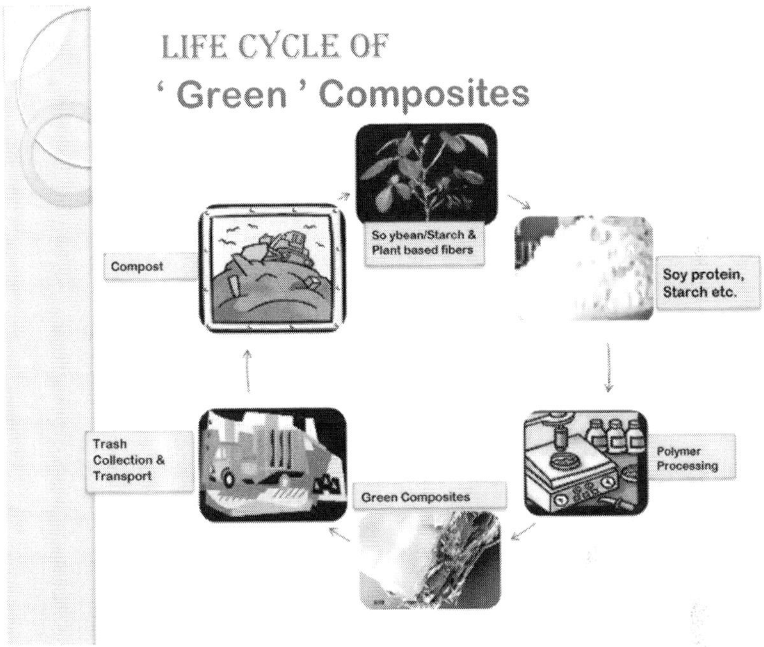

Figure 5.5: Life cycle of green composites.

Source: https://www.slideshare.net/kamblezunjar/green-composites-1-1-2

Energy Information Administration (EIA) stated that the production of oil is predicted to reduce at the rate of four percent per annum after the year of 2010 and consequently excessive effort has been made in the policy and promotion in search of engineering sustainable solutions in energy and the environment.

5.4.1 Constituents of Green Composite

One of the prime element components of the green composites is the application of natural or bio-fiber as strengthening materials. Strengthening material or strengthening are surrounded in the matrix of the compound and create the non-regular or the spread phase of the compound material.

It carries the functional load and stress, and conveys better quality of mechanical properties, which includes stiffness and strength, to the compound. Another important element of green composites are biodegradable polymers which perform as a matrix or regular phase and act as an important role in defining the demanding properties of the green composites.

It offers safety to the compounds in contrast to environmental and chemical attack, and clutches and binds the strengthening materials with each other and transmissions of the load and stress to the strengthening materials.

Reinforcing Bio-Fibers

Bio-fibers are renewable fibers in nature that can be attained from the plants, animals as well as from mineral resources and can be utilized as strengthening material in production of green composites in the same way as the synthetic fibers are made.

In recent years, natural plant-based fibers, such as jute, coconut, hemp, sisal, coir, banana etc., have been utilized as strengthening materials in the place of conventional fibers such as glass, carbon, aramid, etc.

The most important benefits of utilizing these natural fibers over the conventional synthetic fibers are their easy availability, they are cost effective, property of bio-degradability, renewability, recyclability, low density, acceptable specific strength and stiffness, reduced tool wear, non-abrasiveness, easy to produce, light in weight and non-toxic nature.

Plant Fibers

Plants from which natural fibers attained are categorized as primary and secondary according to their use. Primary plants are those plants which grown for their fiber content. The example of plant fiber is jute, hemp, kenaf, and sisal. In addition, the secondary plants, such as pineapple, oil palm and coir. Plant fibers have also been categorized depending on their origin and the part of the plant from which the fiber is pull out.

5.4.2 Biodegradable Polymer Matrix

The yearly disposal of synthetic polymers derivative from products of petroleum in both the United States and European Council country has increased the demands for controlling this non-biodegradable waste. The goods of plastics such as polyethylene, polypropylene, polystyrene and polyvinyl chloride etc. displaced use of products made of metals, glasses, ceramics and wood products more specifically in the field of packaging and these polymers are available in variation of forms as films, bags, containers etc. They persist in the environment; do not undergo degradation after their disposal thus give rise to multitude of ecological and environmental issues. Biodegradable polymers provide the scientists and researchers a possible solution to the problem, not only to the waste disposal issues associated with traditional polymers based on petroleum but also to the increasing in the amount in the price of oil.

5.4.3 PLA Based Green Composites

Different works have been described on the kenaf fiber with polylacticacidPLA. Several reports have been discussed which concerns to chemical enhancement, biodegradability as well as the mechanical properties and dynamic mechanical properties of fabricated compounds.

Silane-treated kenaf fiber strengthened polylacticacidPLA compounds have reported an important enhancement in the mechanical properties.

The heat deflection temperature (HDT) was much higher as compared to the neat PLA. The biodegradability (weight loss with time) of kenaf or polylacticacid PLA reduced nearly about thirty-eight percent after four weeks of decomposition.

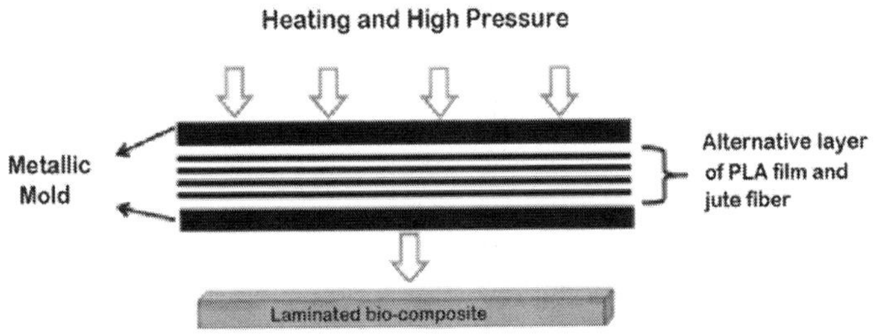

Figure 5.6: Film stacking technique for the production of PLA based green composites.

Source: https://www.researchgate.net/figure/Film-stacking-technique-for-the-development-of-Jute-PLA-based-green-composites_fig2_329884829

Young's modulus (almost six and a half Giga Pascal) and the tensile strength (sixty-two Mega Pascal) of the kenaf (seventy percent) polylacticacid PLA compounds were in the line as compare to conventional compounds.

Flax fiber strengthened polylacticacid PLA compounds was treated and the interfacial characterization was performed using the micro-bond testing method. The interfacial mechanisms were enlightened and discussed based on thermal treatment.

5.4.4 Development Trends of Fully Green Composites

More and more strict environmental rules and guidelines demands for renewable and recyclable materials, the forthcoming energy crisis, the shortage of landfill space, and public issues about the emissions through the course of incineration of traditional Glass fiber reinforced polymer (GFRP) have stimulated efforts to create biodegradable compounds. Governments in nations such as Europe and Japan have encouraged the research and development of fully green composites.

More and more, industries and organizations have attempted to use green composites in their products as well. The most significant example of, Toyota Motor Corporation Limited, which has used green composites made from kenaf fibers to substitute the components made of plastic and metal-based automotive components. Intensive co-operation among industries, research institutes, and governments is important to enhance the development of green composites.

5.5 ARBOFORM AND ITS PROCESSING

The organic polymer lignin is a derivative product of the pulp and paper industry and with each year near about fifty million tons is generated in mills of chemical across the world. Until now, most of the organic polymer has been directly supplied to a thermal application to provide supply of the main energy source in the chemical mills.

The organic polymer lignin is the most important element of a new class of structural engineering and parts for the tools in the use of industries or companies as the only renewable resources.

Arboform is a material of the same group, which consists of lignin, natural fibers for strengthening and natural addons, which help the process and the properties of the polymers. Even though, Arboform it shows properties like wood.

Moreover, Arboform can be processed like a thermoplastic material and can be applied for various products in engineering. The process of blending and amalgamation of granules of this material are based on the standard technologies of engineering of polymers.

Similar to the thermoplastic raw material, they can also be injection molded as well as pressed. The production of the parts of components and the pressed plates from this raw material occurs at lower temperatures.

The resulting parts depict a lower shrinkage as compared to those that are made up of synthetic plastics. However, the mechanical behavior is wood-like and the properties that are measured lie in the range of polyamide.

Human society has utilized and continues to utilize plastic materials because of the cost-effective prices, extensive versatility and the process of manufacturing. All kinds of synthetic plastics are acquired from the petroleum as well as its components.

In addition to that, various plastic materials derived from the resources of fossil are not decomposable. Great use of plastics and various other materials have led to a huge increase of wastes of plastics over the years, which are considered as the municipal solid wastes.

The chief goals of current fabrication as well as future development activities take the following points into account:

- The composite materials acquired from natural resources
- Plastic properties: the component parts and materials of the product which is injected are needed to be provided with the properties of plastic material, verified against the standard test procedures
- Application of standard technologies as well as equipment of processing adjusted by few modifications on the injection moldings
- Recognition of the technical benefits: the materials can be utilized to get high-quality thermoplastics covered with the natural fibers for various use in the consumer product industries, automotive industries, electronic industries, etc.
- Conservation of energy: injection at low temperature that is around 160 °C, thermal stability at around 95° C or even at 105 °C.

Lignin is a natural polymer that is a major component of biomass. Trees and various other woody plants comprise of up to 30% of lignin. This shows the beginning of a modern class of thermoplastics made up of sustainable resources. The increasing interest in the field of green ecology as well as ecological sustainability has contributed to focus the attention on biomass and mainly to the lignocelluloses stock as a renewable, extended and promising resource for biopolymers.

The lignocellulosic biomass consists of three major biopolymers namely cellulose, hemicellulose, and lignin. Now, lignin is considered as an important renewable aromatic resource as it is a great alternative feedstock for the expansion of polymers and chemicals. As an ample bio-polymeric material, cellulose and lignin together represent one of the key components.

ARBOFORM is a wood-based plastic material that can be made along with the production processes of the synthetic industries. Benefits comprise very high tensile strength and very low swelling. As compared to the conventional synthetics, ARBOFORM is known to be heat resistant and shows very low thermic expansion behavior. In upcoming days, lignin free of sulfur could be the basic raw material of this specific product. A suitable method for the decomposition of lignin could be the AQUASOLV-method. The Fraunhofer Institute developed this method for Chemical Technology. ARBOFORM was initially developed in this institute only.

In the year 1996, an injection molding material based upon natural additives, vegetable fibers of hemp, wood or flax and lignin, was successfully manufactured for the very first time. As the basis of raw material provides a modified and disposable alkali lignin of the pulp industry, the modifications of the properties of the material are possible by just changing the derivation, quantity as well as mixing ratio of the substances that are inserted.

For example, tensile strength linearly depends upon the fiber ratio of injection molding formed ARBOFORM. Under the trade name ARBOFORM, this material was improved by just a spin-off of the TECNARO GmbH.

The basic components of materials such as wood fibers and lignin are the by-products of wood and pulp industry that are local as well as regionally accrued. Processing of granulate is possible by extrusion with traditional converting machines of the synthetic industries and by injection molding.

With the 3-zone-screws appropriate for the thermoplastic processing, the granulate will be treated without any other compounds. The temperature of the cylinder lies between 100 °C and 165 °C by processing. The temperature of injector must remain between 155-165 °C. A material based on wood can also be produced by using the method described above.

Commercial uses of the ARBOFORM based on injection molding are loudspeakers, decorative watches, toys or golf tress, etc. Several other products are used in automotive engineering as well as burial branch. Regarding the applications of the material, numerous investigations in the outer sector are taking place.

The granules of ARBOFORM can be processed by various methods, for example, extrusion, injection molding, blow molding, calendaring, pressing or deep drawing into molded parts, semi-finished sheets, products, profiles or films. The frequently used method is injection molding to get ARBOFORM based parts.

The injection molding of the ARBOFORM granules can easily be provided by means of injection equipment. The temperature of injection varies between 155 and 170 °C. One of the nozzles during the process is open 205 to 4.0 mm. the pressure of injection should be so high that it can assure easy rotation of worm gears.

The pressure of injection must as high as possible that are 150 MPa, and so is the speed of injection. The pressure of holding must be around 30% of the capacity of the machine, and the related time is only 0.5-2 seconds. The cooling time should be 20% higher than that of normal thermal plastic material.

The devices of molding injection must be fabricated for the successive production of selected materials. When the component parts of these materials are designed, the devices or instruments available from the injection plastic components of the similar parts might be used for the studies. This kind of operation is continuously providing better outcomes as well as quality.

On the other hand, in complicated cases, the continuous production for the purpose development needs modern instruments particularly adjusted. The following major issues must be considered:

- Shrinkage in the process of molding injection is very low. It should never exceed 0.3%.
- The molds should be fitted along with the mold release spring offering fast operating cycles.
- The low value of shrinkage of Arboform necessitates the use of thin ribs on the molding plates.
- The heated port of nozzles only works in limited situations.

Below are the general guidelines are given for an accurate injection molding processing of the parts of Arboform:

- Granules must not be dried prior to the process. The normal conditions of storage comprise a temperature of 25 °C for drying.
- The manufacturing temperature of the injection molding machine is to be attained by low-melting low-density polyethylene (PE-LD).
- The material should be kept in the machine of injection molding for a maximum of 15 minutes under the conditions of manufacturing temperature. In case of long time of dwell, clean the machine with PE-LD immediately.
- Lignin compounds should not be sustained in the same place with various other hot polymers. As per the process of Arboform manufacturing by using injection molding, can be developed by considering two-level input parameters.

5.6 MECHANICAL PROPERTIES

5.6.1 The Influence of the Fiber Content

The formula of Arboform shows a great dependence of the mechanical features on the fibers content mainly hardwood fibers, which assures huge amount ofvariation in the properties of injection molded parts material. Approximately 45% of content of fibers can give a good balance between strength and stiffness. In case of various other kinds of fibers, the significant high modulus is grasped even when the content of fiber is low.

Basically, the bending stress increases because of the content of fibers. Considering the same percentage, around 45% of the bending stress value is greater than the double value of the tensile strength. With reference to the indentation hardness as well as impact strength in context to the content of fiber, the dependence is very less pronounced.

5.6.2 Micro-indentation

Various researchers showed the results obtained from the micro-indentation of Arboform. The values obtained vary greatly. The tests of micro-indentation were done by the means of a universal device UMT-2 (CETR-Center for Tribology Research, INC, USA). A 2kg sensor was also applied by using the maximum load of 10N.

In a similar manner, the testing performance inferred the application of a Rockwell diamond indenter owing 200-μm radius. The capacity sensor used in testing operations offered measurements of the indenter vertical displacement. The soft use has enabled the reading of values of micro-hardness, values of indentation modulus, and the data of young modulus.

5.6.3 Impact Strength

An enhancement on the processing of compound would lead to a rise of impact strength above 2 kJ/m^2. An additional impact strength growing needs the incorporation of a substantial amount of the impact modifiers that must be biodegradable as well as biocompatible.

On the other hand, a wide range can be covered now. Tensile strength is used to express how much a material can be strained before breaking at high speed and under high stress. Fibers play a crucial role in what concerns the hardness of material or impact strength.

Various studies show that the impact strength majorly reduces the integration of filling materials. By adding 1% of PMDI compatibilizer (polymeric methylene diphenyl diisocyanate) to lignin composites, the impact resistance improves by 92% as compared to that of the non-compatibilizer correspondent.

The process of plasticizing provides this development. Rise of the content of PMDI from 1% to 2% causes a small reduction in the impact strength of composites.

5.7 CARBONIZATION OF HOT-PRESSED ARBO-FORM ®-MIXTURES

Various applications show the strength for porous carbon materials such as filters, electrodes, absorbents or precursors for the penetration with silicon under very high thermal treatment to produce ceramics of carbon and silicon.

Wood pyrolysis, wood plastic composites, wood-based composites has been described earlier, for example by Shafizadeh and Chin (1977), Byrne and Nagle (1997), Kaindl (2000), Sieber et al. (2000), Siegel et al. (2004a), Siegel et al. (2004b), Treusch (2004), Hoffmann (2005), Herzog et al. (2006) and Krischok et al. (2006).

The production of 3-D molded paddings by using solid wood is very limited. The wood carbonization causes an anisotropic shrinkage because of the heterogeneous anatomy of wood. According to several researchers, the shrinkage of beach wood was around 5% to 25% in the axial direction, 15% to 40% in the radial direction and 25% to 40% in the tangential direction.

This process demonstrates the greater risk of cracks and various other damages. After the pyrolysis process and ceramization, a change in the profiles of property is observed. As per Klingler et al. (2003), irretrievable pit closure also reserved silicon infiltration of pine. Hence, panels based on special woods were developed for carbonization as well as ceramization (e.g. Herzog et al. 2006).

Such kind of wood-based panels is homogeneous. Usually, they are categorized by an isotropic shrinkage during the process of carbonization ultimately causing uniform properties of the material. In the case of ceramization, the content of carbon is very significant amongst others.

During the process of carbonization at 1.600 ◦C temperature, the carbon yield of lignin is much higher than the polyoses and cellulose with almost 20%. The content of lignin in beech and pine ranges below 32%. Essentially, great amount of glue with great content of carbon such as phenolic resin were utilized during the process of hot pressing.

In this perspective, an opportunity or chance is the application of wood plastic composites. Such kind of materials consists of wood particles ranging between 50 and 70-wt percentage, mainly uses polypropylene, polyvinylchloride or polyethylene as the material of matrix, and was made by the process of extrusion or injection molding. ARBOFORM is a kind of processable thermoplastic material.

It is manufactured from improved alkali lignin acquired from the natural resources, pulp industries and annual plant fibers such as wood particles and hemp or flax. The Fraunhofer Institute initially developed the material for Chemical Technology in the year 1996 and further TECNARO GmbH, a spin-off of that same institute, advanced it.

Processing is accomplished by the method of extrusion or injection molding. The ranges of density are very high that is around 1300 to 1400 kg/m^3 without using foaming additives. Such kind of traditional products are widespread stereoscopic molds and they cannot be used for any other ceramization.

Another great opportunity is the hot pressing of the plasticized molding material the same as the thermoplastic or curable plastic composite production. It is probable to function in the range of density between 800 kg/m3 and 1400 kg/m3. Usually, the density varies between 800 and 100 kg/m3 for the process of ceramization. ARBOFORM comprises some potential properties, which are appropriate for the successful completion of pyrolysis.

Comparatively, the thermoplastic processable material is isotopic and comprises capillary particles. It has a carbon content of 50-wt percentage. It has the benefit of a huge variety of shapes as compared to the wood-based panels, which are used for ceramization and pyrolysis.

As compared to the solid timber, it is much more homogeneous. Various researchers describe the effects of volume and density reduction as well as mass losses of different kinds of ARBOFORM mixtures. Pyrolysis temperature of 900 °C was opted due to a complete mass loss of phenolic binder and wood caused by the decomposition of organic compounds.

The density of specimen constituted 900 kg/m^3 approximately avoiding bubbling and minor density reduction and providing good infiltration velocity of silicon with optimum silicon carbide yield. The porous carbon material comprises a wide variety of technical uses because of their properties such as mechanical capacity, low-density potential, electric conductivity and resistance to corrosion and heat.

Organic materials such as polyacrylnitrile and phenolic resins are usually applied as a green body for carbon ceramics. Though wood, as well as wood-based composites, are suitable and the hot-pressed ARBOFORM powder mixtures from the wood meal, lignin and natural additives that varied in context to the content of lignin and their additive rates.

The main objective of various researchers is to find out the impact of admixed additives of wood-based composites on the results of carbonization. These researchers because of its homogenous properties as compared to the solid wood also performed several experiments.

Such kinds of production processes are a state of biotechnology and hence, it could be easily possible to develop complex ceramic sand carbon templates that are quite tough to realize by traditional methods of processing. Regarding the thermoplastic process step of ARBOFORM, there are numerous benefits as compared to the wood-based panels.

The sample is pyrolyzed in a nitrogen-containing environment at a temperature of 900 °C. Concerning the absence of cracks, best results can be attained with blend P60 that have a higher content of wood particles. Addition of just 10% of phenolic resin expressively counteracted the reduction of density.

5.8 LIGNIN-BASED COMPOSITES IN PHOTOCATALYSIS

5.8.1 Preparation Techniques of Lignin-Based Composite Photocatalyst

There are several methods used for the production of photocatalyst with lignin as the main support. Immobilizing or depositing a photoactive material within the pores of lignin-based supports may provide significant gains in the Photocatalysis, mainly if the photoactive component can be introduced in an assembled as well as oriented fashion. Also, there are various methods utilized for the manufacturing of lignin-based Photocatalyst like solvent evaporation method, solid-phase synthesis, calcination, pH assisted precipitation and one-pot *in situ* method.

Precipitation is one of the frequently used and cost-effective methods for the production of Photocatalyst composites. More than a layer of photoactive material can be dumped in a homogeneous distribution on the lignin-based support without consuming costly solvents.

Morsella et al. described the production of TiO_2-lignin composite mainly by coating the nanoparticles of TiO_2 with lignin as a shell through pH-assisted precipitation. The process consists of the solubilization of the alkali lignin in an organic solvent (A) and alkaline solvent (B) followed by the addition of TiO_2 particles. Then, sound energy is applied to the final solution to assure homogenization and ultimately TiO_{2-}lignin clusters precipitated out by reducing the pH of the solution by using different acids.

The facile mechano-chemical methods such as ball milling can also be employed for the making of Photocatalyst composites completely based on lignin. For example, lignin-TiO_2 composites were made by wet and dry milling methods in a ball mill. In this specific example, lignin and TiO_2 with a 1:1 mass ratio was milled for 6 hours at 120 rpm.

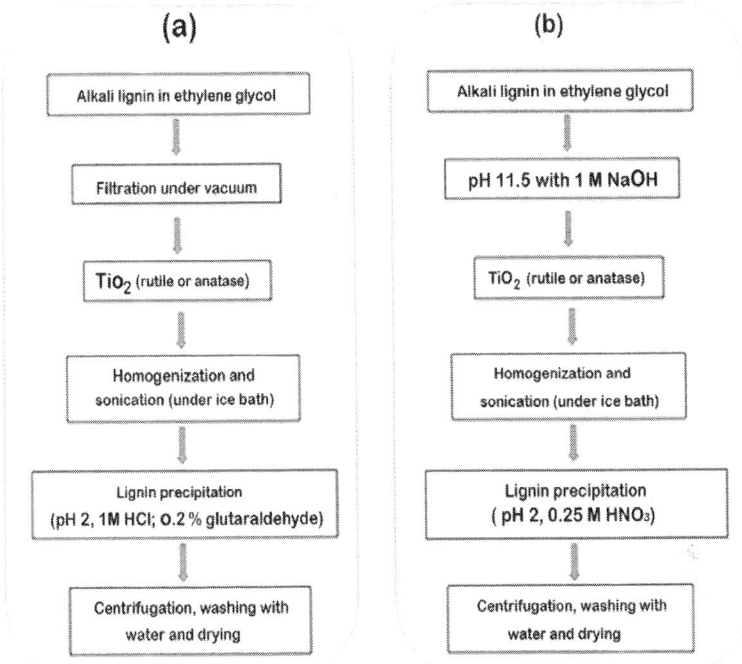

Figure 5.7: Lignin-based composite materials.

Source: https://link.springer.com/article/10.1007/s41061-018-0198-z

The acquired samples are dry milled followed by wet milling by adding various solvents such as water, acetone or hexane with a solvent mass ratio of 2:1. The acquired composites were then filtered and then dried out at 40 °C before using it further.

In addition to that, the one-pot *in situ* method has been utilized for the production of photocatalyst composites. In this particular method, the core photoactive particles are first synthesized, followed by the addition of the template precursor or the coating agent. The reaction mixture is then agitated. The final composite mixture is finally washed, dried and heated to very high temperature in the presence of air or oxygen.

5.8.2 Chemical Interaction between Lignin and Semiconductor

The hydroxyl groups, as well as various other functional groups present in lignin such as carbonyl and carboxyl group, can involve in particular interactions along with the precursors of the functional components like

Photocatalyst and polymers during the production of composites. Currently, lignin has been used as a template to make mesoporous nanoparticles of TiO_2 by using $TiCl_4$ as a precursor.

The electronegative hydroxyl moieties present on the surface of lignin creates a strong affinity towards the metal ions that are electropositive. The positively charged Ti $(OH)_n^{(4-n)+}$ is formed during the process of partial hydrolysis of $TiCl_4$. The Ti $(OH)_n^{(4-n)+}$ has an affinity for nucleophilic ligand, which results in absorption on the lignin surface through electrostatic force of attraction.

After that, the absorbed Ti $(OH)_n^{(4-n)+}$ hydrolyzes and then converts to Ti-(O-lignin)$_4$ all over the surface of lignin. After the completion of hydrolysis, well-dispersed nanoparticles of TiO_2 are made on the lignin surface. Amine groups can easily modify the lignin surface rising the molecular weight and the number of active groups present on the surface.

Production of ZnO nanoparticles by using sodium oxalate ($Na_2C_2O_4$) and zinc nitrate [Zn $(NO_3)_2$], and alkali lignin–amine as a template exposed that the ions that are dissociated from the template of lignin amine at the time of reaction, as well as attractive electrostatic forces, get developed between $C_2O_4^{2-}$ and Zn^{2+} .

Hydrogen bonding, as well as van der Waal forces of attraction, is the main reason behind the direct contact between the particles of ZnO. These forces have been decreased through steric hindrance, which is caused by the template of lignin amine molecule and hence resulted in decreased aggregation and the production of small-sized particles.

5.8.3 Applications of Lignin-Based Composites in Photocatalysis

In the field of material engineering, the interest of using lignin is gradually increasing, particularly in the process of synthesizing composite materials for the photocatalytic applications. Using lignin as a photocatalyst mainly allows controlling the particle size and it aids in obtaining even distribution of the photocatalyst particles.

In addition to that, lignin is obtained from biodegradable biomass as a byproduct from the paper and pulp industries. As a result, the lignin application in material engineering will definitely decrease the cost and it will also aid to fabricate composites that are compatible with the environment. In recent times, several attempts have been made so as to develop photoactive composites based on lignin to enhance the effectiveness of the photolytic process. To overcome the disadvantages of the photolytic reactions, there is a great requirement to understand the fundamental photochemical as well as photo-physical mechanism of the process.

The interface of light and photoactive material is a significant factor in order to determine the effectiveness of the photocatalytic reaction. However, the redox reaction can be inhibited by the huge gap of the band between the conduction band and valence band in the semiconductors.

Thus, UV radiations of high energy are needed for processing the photochemical reaction, which is considered as the basic constraint linked in upgrading a photocatalytic conversion. Lignin usually forms a dark-colored solution in many solvents that makes it inactive. This may also decrease the significance of the product. However, it has also been observed that lignin can be used as a template for the association of hybrid photocatalyst that is stable.

In the case of composites, the interaction between the components plays a major role in recognizing the photocatalytic performance. The features of composites are quite diverse because of the reinforcing synergy between different components of the hybrid photocatalyst. As a result, this improves the efficacy of transfer as well as utilization of the photo-generated electrons. This also intensifies the charge separation that enhances photocatalysis synergistically by overpowering the chances of recombination of electrons and holes.

Thus, the photocatalytic efficiency could easily be enhanced by formulating a composite of photoactive material along with suitable support. Furthermore, the physical and chemical stability of the Photocatalyst mostly depends on the kind of support. Using support soothes the textural properties upon thermal treatment. This improves the dissemination of active sites and generally boosts the catalytic activity of pure oxides.

Photocatalysts that are based on lignin semiconductor composites have huge potential for the curation of contaminated water. Using lignin as photocatalyst support would recommend typical functionality with better physiochemical properties for particular uses endorsed to the interface between support and semiconductor.

Lignin-amine (LA) mesoporous zinc oxide hybrid catalyst showed great sunlight photocatalytic activity. Introduction of amine groups to lignin by means of amination reaction increases the surface activity and the flocculation along with the decolorization efficiency for wastewater treatment.

The temperature of calcination process during the preparation phase of catalyst plays an important role in defining the morphology, size, microstructure as well as photolytic performance of the zinc oxide Nano-photocatalysts. ZnO–LA composite heated at 400 °C shows higher photocatalytic efficiency as compared to those, which are heated at 500 and 600 °C.

The solar photocatalytic act of ZnO–LA is the same as that of TiO_2 (P25). In the course of a photochemical reaction, the water, which is absorbed on the surface, is then oxidized to OH radicals by the holes. On the other hand, the electrons then reduce oxygen, which is absorbed on the surface of ZnO. The superoxide ion and the hydroxyl radical formed result in methyl orange mineralization.

5.9 INNOVATIVE ASPECTS (ENVIRONMENTAL, SOCIAL, ECONOMIC)

ARBOFORM is analogous to the standard synthetic and thermoplastic materials in context to the tensile stress as well as impact strength. It has the same thermal and mechanical properties such as natural wood and therefore associates the positive properties of natural wood with the processing abilities of liquid wood that is thermoplastic materials.

Huge applications of the renewable material will reduce the demand for mineral oil drastically, which is the basic raw material of synthetic thermoplastics. As lignin is a renewable and natural resource with nearly no geographical limitations, it not has the power to avoid issues coming from quantitative as well as geographical limits of this resource but also to decrease the environmental impact of the processing and production of oil.

There is great potential being a thermoplastic material by substituting common synthetic materials like polypropylene and polyethylene is an enormous spectrum of applications. ARBOFORM can be easily processed just as a synthetic thermoplastic material made up of mineral oil. In contrast to the thermoplastic material, ARBOFORM is renewable. Generally, it may decrease the synthetic plastics in any of the existing application.

Around 50 million tons of lignin is produced every year as a byproduct in the paper industry and hence, neither extra harvesting is required nor is it probable that lacks in the supply chain takes place.

REFERENCES

1. ARBOFORM®–liquid wood | Cleaner Production tools & technologies. Retrieved from http://act-clean.eu/index.php/ARBOFORM--liquid-wood;100.218/1
2. Goda, K., & Cao, Y. (2007). Research and Development of Fully Green Composites Reinforced with Natural Fibers. Retrieved from https://www.researchgate.net/publication/237903024_Research_and_Development_of_Fully_Green_Composites_Reinforced_with_Natural_Fibers
3. Jeong, H., Park, J., Kim, S., Lee, J., Ahn, N., & Roh, H. (2013). Preparation and characterization of thermoplastic polyurethanes using partially acetylated kraft lignin. *Fibers and Polymers*, *14*(7), 1082-1093. doi: 10.1007/s12221-013-1082-7
4. Khan, A., Nair, V., Colmenares, J., & Gläser, R. (2018). Lignin-Based Composite Materials for Photocatalysis and Photovoltaics. *Topics In Current Chemistry*, *376*(3). doi: 10.1007/s41061-018-0198-z
5. Plăvănescu, S. (2014). *BIODEGRADABLE COMPOSITE MATERIALS – ARBOFORM: A REVIEW* [Ebook] (p. 22). International Journal of Modern Manufacturing Technologies. Retrieved from http://www.ijmmt.ro/vol6no22014/Plavanescu_Simona.pdf
6. Scholz, G., Lohr, J., Windeisen, E., Tröger, F., & Wegener, G. (2009). Carbonization of hot-pressed ARBOFORM ®-mixtures. *European Journal Of Wood And Wood Products*, *67*(3), 351-355. doi: 10.1007/s00107-009-0347-8
7. Schweizer, M. (2019). *Lignin-based Biocomposites : Applications and new developments*[Ebook] (p. 27). Retrieved from https://zenodo.org/record/32841/files/Lignin-based_biocomposite_for_WPC_profiles_SchweizerTecnaro.pdf
8. Singh, A., Afrin, S., & Karim, Z. (2017). Green Composites: Versatile Material for Future. *Green Biocomposites*, 29-44. doi: 10.1007/978-3-319-49382-4_2

Chapter 6

BIOPLASTICS FROM LIPIDS

CONTENTS

In today's chemical industry, lipids play a crucial role as they are a prominent part of a chemist's synthetic toolbox. This chapter has been framed in such a way that the readers would get an opportunity to have an in-depth understanding about the classifications and functionalities of lipids along with those bacteria that are considered to be the source of lipids in the commercial world. The chapter outlines bacterial lipid producers with respect to lipid accumulation in Gram-positive and Gram-negative bacteria. The Metabolic engineering of lipids have also been described in the chapter.

6.1 INTRODUCTION

In the present day chemical industry, lipids are becoming its major part and are a chief constituent in a synthetic toolbox of an industrial chemist. Numerous different functionalities are produced by the modification of lipids. This, in turn, permits them to produce a huge variety of polymers from a single feedstock.

The whole chapter discusses the definitions of lipids and highlights some of the major regions or sectors in which lipids have been used in the manufacturing of biopolymers till date. An outstanding opportunity is projected by lipids in biopolymer production. Lipids have a tendency of being polymerized directly when some additives are added in the cationic or free-radical process.

The reactivity present in many triglycerides (hydroxyl groups, double bonds) allows for the facile manipulation to whatever the desired functional group for polymerization may be. An outstanding and sustainable chemical for the manufacturing of polymer is triglycerides. Their applications have seen a big boom since the initial times of the 2000s that will continue to grow as concerns over petrochemical feedstocks and prices grow.

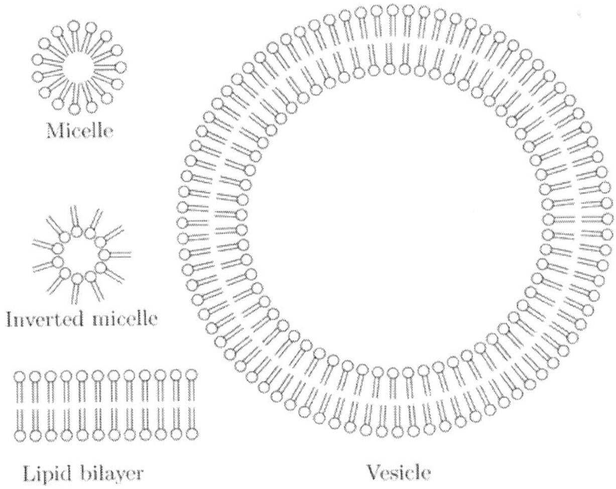

Figure 6.1: The Lipid Vesicle.

Source: http://www.texample.net/media/tikz/examples/PNG/lipid-vesicle.png

6.2 LIPIDS: DEFINITION, CLASSIFICATION AND FUNCTIONS

A heterogeneous group of organic compounds which are major constituent of tissues of animals and plant are termed as lipids. They are randomly grouped together on the basis of their insolubility in water and their solubility in organic solvents like ether, benzene, carbon tetrachloride, and chloroform. The solubility properties of lipids are a function of their alkane-like structures.

Appetizing lipids comprises of an approximate of 25-28 percent of the diet and also aid as a starting material for the manufacturing of many major commodities like soap products. The role played by lipids in our diet is a great matter of concern due to the apparent connection between saturated fats and blood cholesterol with arterial disease. In the diverse kingdom of animals, the major storage compounds are lipids.

On the contrary, plants store a major part of their energy in the form of carbohydrates, of which starch is primary. Apart from serving as energy storage compounds lipids also play many vital roles like maintain the optimum body temperature, providing insulation to the vital organs and shielding them from any mechanical shock. The major constituents of cell membrane structure are lipids and associate with other structures by getting transported across cellular membranes.

6.2.1 Classification of Lipids

Proteins and polysaccharides are polymers whereas lipids are not as they do not have a repeating monomeric unit. But, just like carbohydrates, lipids can be categorized as per their hydrolysis products and on the basis of their similarities in the molecular structure. Following are the three most important subclasses:

Simple lipids:

(a) The hydrolysis of oils and fats that produce glycerol and acids like fatty acids as products.

(b) Hydrolysis of waxes that produces alcohols bearing long-chains and fatty acids as products.

Compound lipids:

- Hydrolysis of phospholipids that produces phosphoric acid, glycerol, fatty acids and an alcohol compound containing nitrogen as products.
- Hydrolysis of glycolipids that produce fatty acids, a carbohydrate and glycerol or sphingosine as products.
- Hydrolysis of sphingolipids produces sphingosine, an alcohol component, phosphoric acid and fatty acids as products.

Steroids:

These are compounds comprising of phenanthrene structure that are quite different from lipids made up of fatty acids.

6.2.2 Physical Properties of Lipids

As discussed before, lipids may either be non-crystalline solids or liquids at room temperature. Opposing widely held belief that pure fats possess properties like tasteless, colorless and odorless, the specific flavors, colors and smell or odor linked with lipids are given to them by external substances that are absorbed by the lipids and are soluble in them. For instance, the pigment carotene is responsible for the yellow color of butter, two compounds formed in the cream as an action of bacteria are 3-hydroxy-2-butanone ($CH_3COCHOOCH_3$) and diacetyl ($CH_3COCOCH_3$) that are responsible for providing taste to butter. The densities of oils and fats are about 0.8 gm/cm^{3,} making them lighter than water. They are an outstanding insulator for the body because they are bad conductors of electricity and heat.

6.2.3 Chemical Properties of Lipids:

Saponification:

There are a number of procedures used for the hydrolysis of triacylglycerols. The most common procedure for the hydrolysis includes the use of an enzyme called lipases or alkali. The process of alkaline hydrolysis is termed as saponification. This is so, because, one of the products of this alkaline hydrolysis is a soap, comprising of potassium or sodium salts of fatty acids.

The aforesaid hydrolysis reaction also serves as an advantageous phenomenon for the foundation of a constant term called saponification number. This constant is a characteristic property of simple lipids. The number of milligrams of potassium hydroxide needed to saponify 1gram of an oil or fat is called the saponification number of a lipid. It indicates the mean molar mass of the lipid.

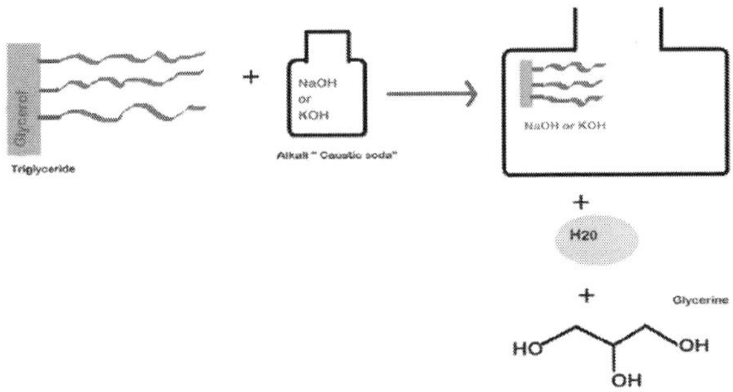

Figure 6.2: Saponification.

Source: https://upload.wikimedia.org/wikipedia/commons/c/c4/Saponification.JPG

A lipid comprising of long-chain fatty acids will account to lesser molecules of acids per unit mass as compared to the one comprising of a short-chain fatty chain. Consequently, a lipid with a preponderance of long-chain fatty acids will have a low saponification number in comparison to a lipid-containing short-chain fatty acids.

It can also be stated as, smaller the saponification number of an oil or fat, higher is the molar mass. The aforesaid point is made clear by the method of determining the saponification number of tristearin. Experimentally a sample of fat is weighed and then saponified with an alcoholic solution of potassium hydroxide which is referred to as a standard solution. After the saponification, the extra alkali is determined by titration with standard acid.

Halogenation:

The reaction by addition of a halogen (such as chlorine, fluorine, bromine or iodine) at the double bond(s) of unsaturated fatty acids, whether they are combined or free as esters in oils or fats is termed as a halogenation reaction. The result of this reaction is observed by the decolorization of the halogen solution.

Due to the consideration that the degree of absorption by an oil or fat is proportional to the number of double bonds present in the fatty acid moieties, the quantity of halogen absorbed by a lipid can be utilized as the index of the degree of unsaturation.

The number of grams of iodine (or iodine equivalent) required to add to 100 grams of oil or fat is termed as iodine number, also called the index value. Numerous factors play a role to influence this value. These factors, for instance, can be degree of unsaturation of each fatty acid and percentage of unsaturated fatty acid in the triacylglycerol molecule.

It is a general fact, that if the iodine number or the index value is high, then the degree of unsaturation will also be high. Natural fats that have a preponderance of saturated fatty acids have iodine numbers lying in the range of 10-50 and those that comprises of plenty of polyunsaturated fatty acids have iodine numbers lying in the range of 120-150.

The determination of the index value or iodine number of triolein is one example. The addition of molecular iodine is represented by the equation. The reagents used in actual practice are interhalogens iodine monobromide (IBr) or iodine monochloride (ICl). Both of the reagents used are more reactive in comparison to iodine alone. A sample of lipid is first weighed and treated with an excess amount of the iodine reagent. Once the reaction is completed, the iodine that was not used i.e. was extra is determined by titration with a standard solution of sodium thiosulfate.

Hydrogenation

The conversion of vegetable oils into solid fats is done commercially on a large scale. Many industries have developed for this purpose. This conversion process from oils to fats by use of hydrogenation is termed as hardening sometimes. Another process comprises of bubbling hydrogen gas under pressure (25 lb./in^2) into a tank of hot oil (200°C) in the presence of finely dispersed nickel as a catalyst.

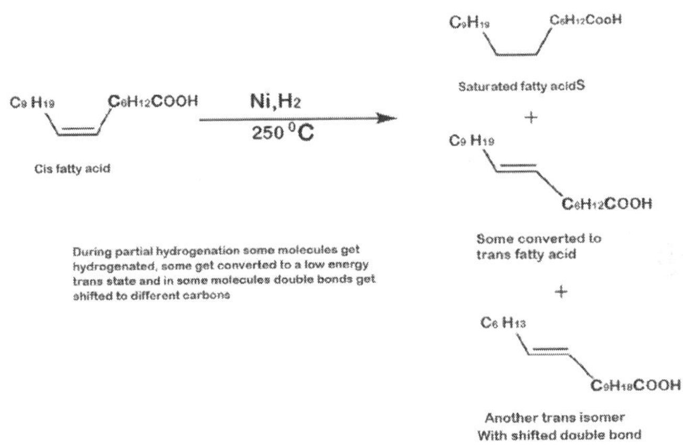

Figure 6.3: Hydrogenation.

Source: https://upload.wikimedia.org/wikipedia/commons/9/94/Hydrogenation.png

6.3 BACTERIA AS SOURCES OF (COMMERCIAL) LIPIDS

6.3.1 Commercial Utilization of Bacterial Lipids

For the purpose of multipurpose commercial applications, lipids became quite valued. They are important also because they have the power to be used in place of fossil resources in numerous industrial processes. A remarkable development has been noticed in the applications of oils as a sustainable replacement feedstock to produce a multitude of basic oleochemicals for organic synthesis, catalysis or biotechnology in the chemical industry.

An important objective around the globe, though, is the alternative for fossil fuels in the sector of transportation and, to date, biodiesel made from plant, animal or microbial oils is the predominant biofuel.

Usually, alcohols with a short chain-like methanol or ethanol along with vegetable oil are used for the transesterification of triacylglycerols (TAG), which produces conventional biodiesel. Since the oil production is extremely poor by some common oil crops, for instance, oil palm yields only about 6000 L/ha and canola about1200 L/ha. This is because the contents of the plant oils are generally less than 5 percent of their biomass.

One more limitation is the dying need for arable land, which no doubt struggles with the world food supply. There are some microorganisms that have quite higher oil contents of roughly about 70 percent of their cell dry weight (CDW). These microorganisms do not require fertile land and only a fractional amount of space for their cultivation. Thus, it is very important to find alternative oil sources in order to substitute fossil fuels, which more sustainable and economically feasible in a long-term way.

It can be attained if the microorganisms used inexpensive industrial by-product, waste compounds, or plant materials that can be generated on rather barren land (termed as "energy crops"). Thus, microorganisms act as a hopeful source for high-value lipids and offer many advantages over conventionally used oils with animal or plant origin.

The metabolism, genome and cell compartmentation are low in bacteria and are capable of producing a huge range of various kinds of lipids as compared to multicellular eukaryotes. Also, they can be readily modified genetically in order to obtain optimized strains which further will increase the competitiveness and productivity of the entire process.

Wax esters (WE) are another kind of lipids that comprises long chain alcohol and fatty acid residues. This is generally not accumulated by eukaryotes in significant amounts. Wax esters are rich ingredients in various products such as lubricants or coatings, cosmetic or medical products, amongst others.

Generally, they are harvested using plants like carnauba wax or jojoba. This requires expensive cultivation and a time-consuming process. Specific bacteria are able to provide a good substitute source of wax esters (WE) with an identical composition as compared to jojoba oil. Later, the preferred composition can be targeted by the choice of precursor substrates.

Figure 6.4: Carnauba wax texture.

Source: https://cdn.pixabay.com/photo/2016/05/19/14/42/carnauba-wax-1403125_960_720.jpg

On the basis of the current knowledge, noteworthy bacterial TAG or WE synthesis is catalyzed by the wax ester synthase/acyl-CoA: diacylglycerol acyltransferase (WS/DGAT). Though there exists the alternative source for TAG synthesis in some bacterial species, they generally do not account for a significant accumulation of this storage compound.

6.3.2 Feedstock for the Production of Bacterial Lipids

The use of starch-derived sugars or sucrose for the bacterial production of lipids on a large scale is problematic both ethically and economically. Generally, a large variety of lipids can be produced efficiently from substrates that are structurally related such as alkanoic acid and alkanes.

However, the economic production with these substrates is quite problematic and unlikely because of several reasons like their low miscibility and toxicity, respectively, and their high cost. Since the substrate price can donate to about 50 percent of the cost of the total production, it affects the price competitiveness of lipid-derived fuels greatly as compared to conventional fuels.

Therefore, in order to minimize the production, it is inevitable to engage low priced carbon and/or nitrogen sources from industrial or agricultural waste and surplus material, for instance, sugar-rich molasses, hydrolyzed plant biomass, glycerol from biodiesel production whey as complex nitrogen and phosphorus source.

Lignocellulose from the forest, agricultural and industrial residuals worldwide denote the cheapest and largest resource of sugars with near about 4.15 billion tons of different kinds of waste from agriculture accruing annually. By the use of this low priced and surplus feedstock, the production of lipids will not only become ecologically stable but also geopolitically independent from oil-producing countries.

For the reason of the stable and complex structures of lignocellulose, a lot of time and energy is consumed pretreatment during the breakdown into fermentable sugars. Thus, it will be beneficial to engage bacterial strains either degrade (hemi) cellulose or powerfully utilize the evolved hexoses and pentoses, for instance, D-galactose, L-arabinose, D-mannose, D-xylose and, D-glucose in order to generate "second-generation biofuels"

Other thought-provoking methodologies that are discussed in recent times target at creating processes in effect, in order to attain microbial lipids formed using carbon dioxide or organic waste for the manufacturing of "third-generation biofuels". During wastewater treatment, specialized anaerobic microorganisms accumulate lipid mixtures composed of TAG, WE or PHA.

It has been estimated that transesterification and extraction of these lipids in 50 percent of the total biological wastewater treatment plants belonging to the U.S. could make up 0.5 percent of its annual demand for diesel and hence capable of playing a role in strategies for future energy that utilizes the possible number of renewable resources.

Oleaginous photosynthetic microorganisms are capable of recycling carbon dioxide and converting solar energy into lipids. Some cyanobacteria accrue a major amount of lipids in the form of diacylglycerol in their thylakoid membranes and in order to attain overproduce of FFA it can be genetically engineered whereas; under environmental stress, TAG is stored by microalgae.

6.4 BACTERIAL LIPID PRODUCERS

While evaluating the bacterial strains concerning a powerful application as industrial lipid producer, many major factors should be taken into consideration. These factors include:

- Media supplements and high-priced carbon sources should not be essential.
- The rate of development, product concentrations and attainable density of cell, specifically under conditions of fermentation, should be high
- Product extraction should be as easy as possible. Continuous cultivation and reduced cost of extraction are enabled by extracellular product.

Furthermore, the bacteria must be non-pathogenic and genetically accessible if directed strain optimization is necessary.

6.4.1 Lipid Accumulation in Gram-positive Bacteria

The capability to accumulate major amounts of TAG as intracellular storage deposits is prominently discovered in bacteria that belong to the Gram-positive Actinomycetales, for example, *Gordonia*, *Rhodococcus* or *Streptomyces* species, *Dietzia*, *Arthrobacter*. Thus specifically *R. opacus* PD630 is broadly known for its remarkable TAG contents up around 80 percent of its CDW and observed as a potential strain for industrial production of TAG because it grows rapidly, accepts a huge variety of carbon sources and high substrate tolerance.

Though many years ago it was reported that *Arthrobacter* AK19 can accumulate higher amounts similarly, but still this strain was not utilized for further examinations. Gordonia or Rhodococcus strains specifically seem suitable for the lignocellulose hydrolysates or conversion of industrial waste like molasses, dairy wastewater, wastes generated from agro-industrial processes. As compared to this, *Streptomyces* isolate can use cellobiose to produce up to 50 percent fatty acids of its CDW. Streptomycetes have a unique characteristic to also include branched chain fatty acids, whereas Rhodococcus used for the preparation of TAG, generally comprises of straight-chain C16 – C18 fatty acids residues.

It is to be noted that contents of lipids are usually measured in batch cultivation, through continuous cultivation enables a major rise in the productivity of lipids overall. This does not mean that the lipids with high cellular content would automatically make the process achievable economically, as the yield of the lipid overall might be low comparatively. For example, the content of lipid after batch cultivation of *R. opacus* cells along with molasses was observed to be 92 percent of its CDW, only around 55 mg/L, corresponding to a productivity of approx. 0.6 mg/L·h were reached. Contrarily, fed – batch fermentation of 30 liters the productivity of *R. opacus* saw a major rise to 19.4 g/L or 0.38 g/L·h, respectively, of total fatty acids (corresponding to approx. 50% of its CDW) from molasses and sucrose. Therefore, the attained lipid yield and individual cultivations are not certainly comparable.

6.4.2 Lipid Accumulation in Gram-negative Bacteria

The amount of TAG storage as attained by some Gram-positive bacteria is not reported to be achieved by Gram-negative bacteria. Rather, some Gram-negative bacteria, in specifically hydrocarbon clastic bacteria or Acinetobacter, like *Marinobacter* species or Alcanivorax, have the capability of forming WE, which are later accumulated with less amount of TAG in mixed lipid incorporations. Significant amounts are generally attained when the same bacteria are cultivated with relatable carbon sources, such as olive oil or n- alkanes.

An interesting Gram-negative isolate, *Aeromonas* sp. 3010, attained a total lipid content of approximately 12 percent, whereof up to 30% is eicosapentaenoic acid [20:5 n-3, EPA], a polyunsaturated fatty acid with great importance for food or pharmaceutical industries.

Moreover, two unusual discoveries of the Gram-negative bacteria that produces lipids have been stated.

An isolate allotted to the genus *Nitratireductor* (an α-proteobacterium), that could be fascinating for waste-water treatment, develops with short-chain organic acids and produces an unfamiliar mixture of lipids made up of squalene, TAG and light oils (such as 2-butenoic acid methyl ester) with fillings of up to 70% of the CDW [57]. in addition to this, a chemolithotrophic, CO_2-sequestering *Serratia* strain produced nearly 50 percent of its CDW as lipids and long-chain hydrocarbons, and up to 466 mg extracellular lipids/L, respectively. However, reports exist based on Bacillus and oleaginous Psuedomonas strains, which need validation as it is doubtful that aforesaid species have the capability of accumulating neutral lipids rather are known because of their capability to store PHA.

These isolates might not have been correctly assigned at the taxonomic level. Moreover, the existence of neutral lipids in *P. aeruginosa* 44T1 was not established in a sovereign study. Another unusual case is denoted by cyanobacteria, due to the intracellular TAG storage yet has not been established in this category of microorganisms. Presence of (WS/) DGAT gene is not detected in published cyanobacterial genomes. On the other hand, in diacylglycerols and membrane lipids, photosynthetic membranes act as sources of large storage tanks, that can be used biotechnologically. For example, for transesterification with methanol to yield biodiesel.

By using carbon dioxide alone as a source of carbon, some cyanobacterial strains have made up significant lipid content of greater than 40 percent of the CDW, for example *Synechococcus* or *Cyanobacterium aponinum*, but lipids are complex compounds that require hydrolysis to evolve the needed fatty acids.

For the manufacturing of lipids on a large scale such as feedstock for biodiesel, specifically bacteria belonging to the actinomycetes, like certain *Streptomyces* species or Rhodococcus, denote healthy productions strains which have the ability to develop on low priced raw materials and can accumulate large amounts of TAG.

Though Gram-negative bacteria used for the synthesis of WE generally are not able to attain same kind lipid with high content and needs much higher priced sources of carbon, some WEs with required composition can be manufactured for instead end products with high-value for example, cosmetics.

6.5 BIOSYNTHESIS OF LIPIDS

Lipids act as a source for a number of cellular roles which includes some which are recognized in recent times. Lipids act as the major form of stored energy in most organisms. They also act as a principal constituent of cell membranes. Specific lipids act as detergents (bile salts), pigments (retinal), hormones (vitamin D derivatives, sex hormones), and cofactors (vitamin K), anchors for membrane proteins (covalently attached fatty acids, prenyl groups, and phosphatidylinositol), extracellular and intracellular messengers (eicosanoids and derivatives of phosphatidylinositol) and transporters (dolichols).

Therefore, it is important to all organisms to have the capability of producing different types of lipids. The sequences of these reactions like other biosynthetic pathways are reductive as well as endergonic. The source of energy used by them is ATP (Adenosine triphosphate) and the reductant used is reduced electron carrier (generally NADPH).

6.5.1 Malonyl and Acetyl Groups are received by Fatty Acid Synthase before the Condensation Part

Chemical reactions that are responsible for the building of the Fatty Acids can start the two thiol groups on the complex enzyme structure and must be charged with appropriate Acyl group. The first step is the synthase, that is, transferring of the acetyl group of acetyls- CoA to the Cys -SH group of the β- ketoacyl-ACP. The catalyst used in this reaction is the acetyl- CoA - ACP transacetylase. The second step is the transferring of the malonyl group from malonyl-CoA to the -SH group of ACP. The catalyst used in this reaction is the malonyl-CoA-ACP transferase, which is also a part of the complex used. In the charged synthase complex mentioned above, the malonyl and acetyl groups are found to be very close to one another and are readily active for the process of chain lengthening that includes the four steps that are described in the earlier section. To understand the process in detail, these steps are further explained in detail below:

- Condensation: This is the first step in the formation of a fatty acid chain. Condensation of the activated malonyl groups and acetyl groups takes place in order to form acetoacetyl group that is bound to ACP via the phosphopantetheine -SH group: acetoacetyl-ACP. In conjunction, a molecule of COZ is developed simultaneously. The catalyst in this reaction used is β-ketoacyl-ACP synthase. The catalyst works as the transfer of acetyl group takes place from the Cys – SH group of the enzyme to the malonyl group on the –SH of ACP, that gives rise to a new acetoacetyl unit, called methyl-terminal two-carbon unit.

Further, it can be noted that the carbon atom formed in CO_2 during this reaction is the same carbon atom introduced initially by the acetyl-CoA carboxylase reaction (by reacting malonyl-CoA with HCO3). Hence, carbon dioxide is only briefly in-covalent linking at the time of biosynthesis of fatty acids, and hence it is removed whenever a two-carbon unit is added.

The question that arises here is that, "Why do cells go to the trouble of adding CO_2 to make a malonyl group from an acetyl group, only to lose CO_2 again during the formation of acetoacetate?"

For this, it must be kept under consideration that in the β oxidation of fatty acids, linkage of the two acyl groups bond i.e. linkage between the acyl chain and the acetyl unit is very exergonic in nature. Hence, a process of simple condensation of two acyl groups, for example, two acetyl CoA molecule is highly endergonic in nature.

Further, by substituting the activated malonyl, in place of acetyl groups, the condensation reaction is given a bend toward more thermodynamic favorable conditions. The methylene carbon or C-2 of the malonyl group, clubbed between carboxyl carbons and carbonyl, is particularly a preferred nucleophile. Additionally, in the condensation process, the enzyme's –SH group is displacing, by the decarboxylation of the malonyl group. This, in turn, promotes the nucleophilic attack of this methylene carbon at the thioester that links the β-ketoacyl-ACP synthase and acetyl group.

Linking of the condensation and decarboxylation of the malonyl group makes the whole process very exergonic in nature. As discussed earlier, it is similar to the carboxylation decarboxylation sequence that enables the formation of phosphoenolpyruvate from the pyruvate in gluconeogenesis.

With the help of activated malonyl groups in the production of fatty acids and using the activated acetate in the degradation of fatty acids, the cells were able to manage both processes favorably, as the one process is exactly the reverse of the other. Also, the ATP, which is used to produce malonyl-CoA from acetyl-CoA and HCO_3^{--}, is used to provide the extra energy required to make the synthesis of fatty acid much favorable.

- Reduction of the Carbonyl Group: The Acetoacetyl- ACP developed during the condensation phase further undergoes for the process of reduction of the carbonyl group at C-3 to form D-β-hydroxy butyryl-ACP. The catalyst used in this reaction is the /3-ketoacyl-ACP/ reductase and NADPH as an electron donor in this reaction. It can also be noted that the stereoisomeric forms of both the D-β-hydroxy butyryl group and the L-β-hydroxyaryl as it acts as an intermediate in the fatty acid oxidation.
- Dehydration: this is the third step and in this stage, the water molecules are removed from C-3 and C-2 of D-β-hydroxy butyryl-ACP in order to produce a double bond in the compound, trans-Δ2-butenoyl-ACP. The catalyst used in this dehydration process is i-hydroxyaryl-ACP dehydratase enzyme.
- Reduction of the Double Bond: this is the final step where the reduction (Saturated) of the double bond of trans-Δ2-butenoyl-ACP takes place. This reaction creates butyryl-ACP, with the help of catalyst (enzyme) enoyl- ACP reductase. In this reaction also, the NADPH acts as an electron donor.

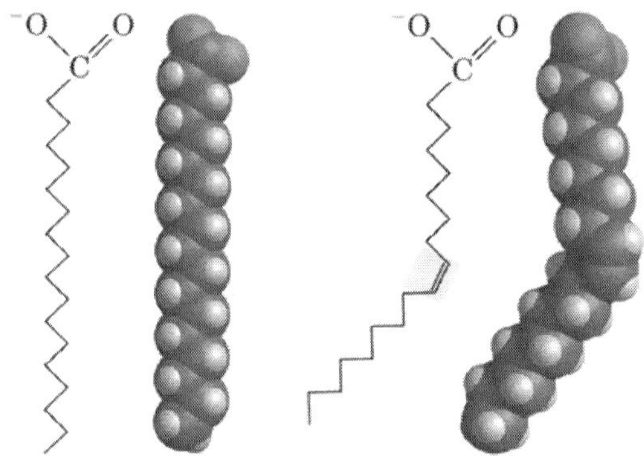

Figure 6.5: Lipids and fatty acids as saturated and unsaturated fatty acids.

Source: https://farm7.staticflickr.com/6184/6082438756_9eff621428.jpg

6.6 METABOLIC ENGINEERING OF LIPIDS

Oils extracted from plants are useful and about two-third of them are used as food and rest one third are utilized for industrial purposes. There are numerous applications of plant oils in industries like in the manufacturing of detergents, biofuels, soaps, and lubricants. The other compounds that are used for the production of oils include cytoplasm, endoplasmic reticulum, and plastids.

Oils are stored in the lipid bodies post-synthesis. Acetate molecules are extracted from the cytoplasm by the plastids and in converted into malonyl CoA and acetyl CoA. A series of reactions are performed on these two molecules and then the Fatty Acid Synthase or FAS is used as a catalyst to obtain the Fatty Acyl Carrier proteins that have different carbon atoms.

With the action of Acyl – ACP thioesterases, the fatty acids are released. These fatty acids are further exported to the cytoplasm. During this phase, these fatty acids may undergo some modifications like desaturation (insertion of double bonds), elongation and hydroxylation, etc.

These altered fatty acids then come in contact with glycerol 3-phosphate that is present in the endoplasmic reticulum and the final product obtained is the Triacylglycerol or TG. These TG are then taken to lipid bodies and are kept there. In the lipid monolayers, the oil bodies of seed contain Oleosins, which is a type of protein. To perform genetic engineering to manipulate the plant oil biosynthesis, some of the approaches are described below in brief:

6.6.1 Production of Shorter Chain Fatty Acids

Most of the plant oils have greater than 16 carbon atoms, for example, stearic, palmitic and oleic acids. Oils that have shorter chains (C_8—C_{14}) are preferred for the uses like for the production of detergents, soaps, cosmetics etc. The hydrolysis of acyl- ACP can possibly be terminated by using some particular thioesterases. This will help in producing high levels of fatty acids that are best suited for industrial purposes. California Bay Tree or *Umbellularia californica* releases acyl-ACP thioesterase that is specifically used in hydrolyzes of lauroyl ACP. With scientific advancements, the gene encoding of this enzyme has been successfully cloned and is transferred to oilseed rape. These transgenic plants are known for producing oils with a high percentage of lauric acid (12-carbon fatty acid).

6.6.2 Producing Long Chain Fatty Acids

In the industrial applications of oils, longer chain fatty acids ($>C_{18}$) like triacylglycerol's are used. Hence, Erucic Acid with 22 carbon atoms, containing oils is used in industrial applications. Although, Erucic acid is not recommended for human consumption. With the traditional plant breeding, two different crops of oilseed rape have been created namely - low-erucic acid rape (LEAR) and high-erucic acid rape (HEAR).

There has been a constant practice and research going on to genetically modify the crops so that they will yield oils with a high amount of erucic acid. The methodology includes the overexpressing of the gene encoding the enzymes elongates and transfer of genes to produce enzymes that can preferentially incorporate erucic acid into triacylglycerols.

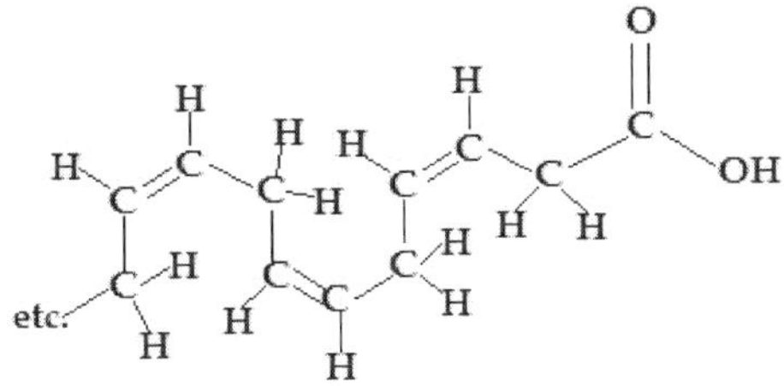

Figure 6.6: Fatty Acid Chain.

Source: https://www.flickr.com/photos/98593409@N05/9299866904

6.6.3 Producing Fatty Acids by the help of Modified Degree of Saturation

Any Fatty acid with no double bonds is said to be saturated Fatty Acids. Unsaturated and Saturated (with one or more double bonds) are found in nature. The genetic manipulation of degree of saturation in fatty acids can be is possible with the latest technological advancements.

Production of unsaturated fatty acids:

Oleic acid (C_{18}: 1) rich oils are used as feed, food and in some industries. There has been some success in the transfer of antisense gene encoding the enzyme desaturase in soybean plant and oilseed rape. These transgenic plants were found to give oils with a very high amount of oleic acid.

Production of saturated fatty acids:

At times, it is required to produce oils with high contents of saturated fatty acids. Success has been reported by using of different approaches like antisense RNA approach. The aim was to minimize the conversion of stearate (C_{18}: 0) to unsaturated fatty acids such as linoleic acid (C_{1818}: 2), oleic acid (C_{18}: 1) and linolenic acid (C_{18}: 3).

The enzyme stearoyl-ACP desaturase is used as a catalyst for the conversion of oleoyl-ACP from stearoyl-ACP. The gene encoding the enzyme stearoyl-ACP desaturase has been isolated from Brassica rape and its complementary sequence cloned. The transgenic plants created by this approach were found to contain a high amount of saturated fatty acid namely stearic acid, and low concentration of oleic acid.

6.7 VEGETABLE OIL-BASED POLYMERS: SYNTHESIS AND APPLICATIONS

There are several reasons behind conducting extensive research on the aspects of the production of renewable thermosetting polymers. These polymers are produced from plant seed oils and have been helpful in diverting researchers' mind from the petrochemical-based polymers. The reason behind such polymers are:

- Deteriorating health issues;
- Rigorous environmental protection policies,

- An exhaustive search not-so expensive and alternative materials, and
- The pursuit of renewability, sustainability, and high-performance materials for technical applications.

It is the Bio-based polymers that have accumulated a lot of appreciation and interest amongst the researchers. This is so because they are good for health, they are environmentally friendly as they are associated with conventional synthetic polymers. Also, these polymers are biodegradable in nature, can be easily renewed and are environmentally-friendly.

The most prominent example for the bio-based polymers could be of the thermosetting polymers. These polymers are produced from plant seeds like sunflower oil, linseed oil, soybean oil and cashew nut oil. Apart from this, there are certain thermoplastic counterparts like PLA which are produced from corn or maize. Besides, there is poly-hydroxybutyrate, poly caprolactone etc. The main problem behind the production of these counterparts is the cost required to be invested in them. But it is important to note that their production has become a viable proposition as technologies have evolved.

Figure 6.7: Soybean Oil.

Source: https://farm3.staticflickr.com/2888/10059732523_0579962b96_b.jpg

Bio-based polymers are used for multiple purposes. For example, they are used for sustainable alternatives to traditional petrochemical-derived materials. Some of those materials are phenol-formaldehyde, epoxy resin, unsaturated polyester resin, polyurethane, phenolic resin, and iso-cyanate resin in the manufacture of composite and in coating applications.

6.8 LIPID-BASED EDIBLE FILMS

There are several petrochemical polymers that are also known as plastics. These plastics are generally used in the field of food packaging. Also, these polymers are easy to use and are highly cost-effective. But, like everything else, polymers also have a big drawback and that is that they are not good for health and are harmful to the environment. Regarding health problems, polymeric materials are not inert.

Figure 6.8: Plastics used for food packaging.

Source: https://farm8.staticflickr.com/7140/7460449608_e7027833c2_b.jpg

It is mostly dependent upon the food components and the ingredients that are used to produce the food product which will let the chemicals from the polymers get transferred to them thereby causing problems to the health of an individual. In case of the composites, fat has the strongest influence. This is so because it increases the mobility of plastic film ingredients which further increases the migration of plastic molecules into food. Due to these the characteristics, packaging material may get altered.

To make it easier to understand, it can be stated that the when comparing fat with water, it has been observed that the polymers have higher level of solubility of migrating organic compounds in fat. This is more of a concern about health while in case of environmental policies, there are problems with recycling and incineration. In order to overcome these two problems, edible polymers with food-grade additives and biodegradable packaging polymers should be promoted.

Such polymers are further distributed into 4 different groups. Those four groups are:

- Polysaccharides,
- Proteins,
- Lipids and
- Complex.

Among these four groups, there are two groups that have high mechanical and optical properties. Those two groups of polymers are polysaccharides and proteins. But these two groups of polymers are highly sensitive to moisture.

6.8.1 Application to Fruits and Vegetables

With the developments taking place in almost all the sectors, the needs of the consumers have also elevated significantly. People are not restricted to just local products, but are highly inclined towards the imported products and are ready to spend more on them. But there is always a problem associated with such imported products, and also the problem of packaging due to different transport conditions.

Therefore, to elevate the life of the packaged food products which are desired by the consumers, several traditional methodologies like drying are now considered to keep the products fresh and stain free. Since most of the fruits and vegetables contain more than 80-90% of water, it is important to preserve that amount of water.

There shouldn't be a water vapor transfer. Fruits and vegetables stored for a long time can be coated with different formulations including paraffin and natural waxes. Additionally, some chemical substances like fungicides or growth regulators can be incorporated.

Yet, before distribution to markets and processing these constituents should be removed. This can be done by rinsing or cleaning. Respiration process is another problem beyond water loss since it causes carbohydrates to produce carbon dioxide, water and heat, which will, in turn, lead to a loss in carbohydrate content. Moreover, it can be seen in their organoleptic properties.

Figure 6.9: Food Packaging.

Source: https://www.flickr.com/photos/stone-soup/5007044417/

There will be changes in their color, taste and odor. Some researches proved that incorporation of beeswax into chitosan-based coatings caused a remarkable decrease in the respiration rate of strawberries. Coatings incorporated with lipids are generally a good solution owing to their stability.

However, they can also negatively affect the sensory parameters, leading a waxy sensation. That's why there are still some studies trying to create composite edible films and coatings.

6.8.2 Application to Dairy Products

Among all the food products, it has been observed that cheese is one most diversified and versatile product which can be easily used and blended with almost every dish. Generally, cheese is produced through starter lactic acid bacteria which is also known as SLAB but not always. The shelf-life of cheese which are produced by using SLAB has a little bit longer shelf-life comparing to those not produced by using SLAB.

Yet, in any way, shelf life of most cheese types are limited due to uncontrolled fungal and bacterial development on the surface. Therefore, there are several edible films available in the market that can help in reducing the negative impact of all such changes.

Ramos et al. (2012) used whey protein isolate, glycerol, guar gum, sunflower oil and Tween 20 with several combinations of antimicrobial compounds like natamycin, lactic acid. After experimenting and bringing all these ingredients together, it was concluded that the application of these coatings restricts contaminant microorganism and reduces water loss during 60 days of storage. In another study, galactomannan coatings which are plasticized using glycerol containing corn oil reduced O_2 consumption and CO_2 production rates on semi-hard regional cheese.

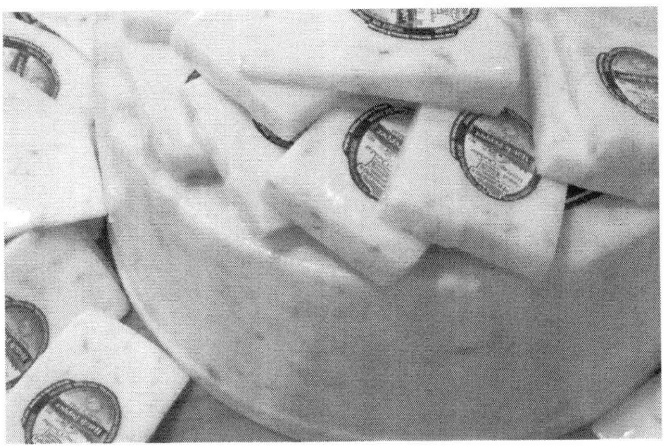

Figure 6.10: Cheese produced by using SLAB.

Source: https://pxhere.com/en/photo/953206

6.8.3 Application in Bakery Products

The first thing that comes in the mind when a person hears about bakery product is the crisp and the crunch of the products. Products like, biscuits, cereals, oatmeals have crispiness with an added moisture content. When the bakery products are not crisp and have soft texture than it must be understood that they have increased water content.

The water transfer doesn't always occur from the surrounding atmosphere to food product but also through the food products. Dried fruits have higher water activity comparing to cereals.

Therefore, it is important to have proper water transfer. Talens, Perez-Masia, Fabra, Vargas, and Chiralt (2012) are some of the famous researchers who studied the effectiveness of edible coatings based on caseinates and chitosan in partially dried pineapple for its use in dry fruit-cereal products. According to them, they came to the conclusion that caseinate-based emulsion coatings prolonged the shelf life of the product.

Figure 6.11: Crispy Bakery Products.

Source: https://www.maxpixel.net/Baked-Goods-Roll-Bakery-Crispy-Pastries-Bake-3286363

REFERENCES

1. Aydin, Furkan & Kahve, Halil & Ardic, Mustafa. (2017). Lipid-based edible films. The Journal of Scientific and Engineering Research. 4. 86-92.
2. Bioinfo.org.cn. (n.d.). *Chapter 20 : Lipid Biosynthesis*. [online] Available at: http://www.bioinfo.org.cn/book/biochemistry/chapt20/bio1.htm [Accessed 29 Mar. 2019].
3. Biology Discussion. (n.d.). *Metabolic Engineering of Lipids: A Close View*. [online] Available at: http://www.biologydiscussion.com/genetics/engineering/metabolic-engineering-of-lipids-a-close-view/10864 [Accessed 29 Mar. 2019].
4. Coles, S. (2013). Bioplastics from Lipids. *Bio-Based Plastics*, [online] pp.117-135. Available at: https://onlinelibrary.wiley.com/doi/10.1002/9781118676646.ch6 [Accessed 29 Mar. 2019].
5. Lligadas, G., Ronda, J., Galià, M. and Cádiz, V. (2013). Renewable polymeric materials from vegetable oils: a perspective. *Materials Today*, [online] 16(9), pp.337-343. Available at: https://www.sciencedirect.com/science/article/pii/S1369702113002940 [Accessed 29 Mar. 2019].
6. Lligadas, G., Ronda, J., Galià, M. and Cádiz, V. (2013). Renewable polymeric materials from vegetable oils: a perspective. *Materials Today*, [online] 16(9), pp.337-343. Available at: https://www.sciencedirect.com/science/article/pii/S1369702113002940 [Accessed 29 Mar. 2019].
7. Zoology Notes. (n.d.). *Lipids: Definition, Classification an Functions | Biochemistry*. [online] Available at: http://www.notesonzoology.com/lipids/lipids-definition-classification-an-functions-biochemistry/3510 [Accessed 29 Mar. 2019].

Chapter 7

LACTIC ACIDS AND CRYSTALLIZATION FOR BIO-BASED PLASTICS

CONTENTS

Bio plastics are a new trend in the manufacturing industry and particularly that which caters to the production of plastic and its products. In the day to day lives, people are surrounded by plastic products such as plastic utensils, food containers, boxes, water bottles and other such products. Though, recently, researches have found that the conventional plastic products, are neither good for health nor favorable for the environment as they take a lot of time to degrade and get back into the nature, that too causing harm to the environment by releasing substances and gases that may damage the ecosystem. Here, the use of bio plastics comes into the picture and most widely used substance to manufacture bioplastics, which is gaining attention increasingly, is lactic acid. PLA, which will be informed about later in the chapter, is a key member of the bio plastics community. This chapter centers around the use of PLA and crystallization process in the manufacturing of bio plastics and based products and enlists the various uses of such products.

7.1 INTRODUCTION

The development of the new resulting materials out from the renewable sources is the main goal which a has high priority on a technological and environmental point of view to decrease the dependency on the non-renewable source such as oil.

In the present time, the utilization of products made from plastic is more than two hundred million tons all across the globe. The consumption of plastic products accounted for an increment of nearly about five percent twelve-monthly. Till the present time, plastic products made from petrochemicals have been used in the form of material for packaging. This type of plastics which are made with the help of Petro-chemical are preferred because these materials have good properties such as mechanical properties, heat stability, good barriers properties, easily available, and are relatively cost-efficient.

Never the less, the end-life management of these plastics is not an easy task, because only some of the plastic products, for example, those produced of only one material, can be selectively collected and recycled. Films of plastic can be recycled and reused but they have a low apparent density. The packages which contain several numbers of materials, for example, packages with multi-layer protection, the process of recycling is much difficult, but they can neither be composted.

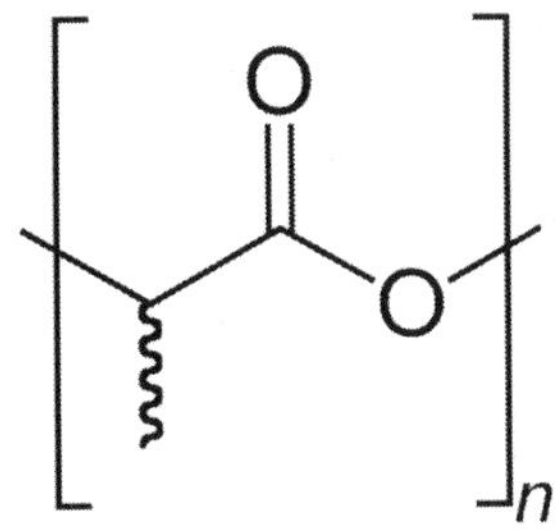

Figure 7.1: Skeletal representation of Polylactic acid.

Source: https://upload.wikimedia.org/wikipedia/commons/thumb/7/73/Polylactic-acid-2D-skeletal.png/495px-Polylactic-acid-2D-skeletal.png

The making of flexible films with the help of plastic which can be composted can be specifically crucial. They can be applied to make low apparent-density packaging, for example, bags or pouches, or packaging with multi-layer protection though coupling them with board or paper.

And consequently, fully composted packaging is manufactured. Although, for addressing this market, it is more important to find a solution to this problem of low tear resistance of this type of composted films. The solution to these issues will permit their process of making with the help of various approaches of traditional industrial processing. For example, blown film extrusion.

Polylacticacid is a polymer which does not impact the environment because polylacticacid is compostable in nature. Polylacticacid is derived from renewable resources. Polylacticacid demonstrates high modulus of elasticity and has high transparency. Polylacticacid has also strong limitation which is low glass transition temperature, low thermal stability, high brightness and rate of crystallization is low.

The making of polylacticacid demonstrates several numbers of benefits because it is renewable in nature. In addition, the making of polylacticacid utilizes more amount of carbon dioxide. Polylacticacid offers considerable energy saving.

Polylacticacid can be recycled, and it is compostable in nature. The physical properties and mechanical properties can be partially personalized by properly selecting the polymer architecture. In 2017, polylacticacidthe polylactic acid (PLA) market globally is valued at USD 698.27 million, by revenue (Mordor Intelligence).

. At present, polylacticacid represents one of the cost-effective bio-based and biodegradable substitute to plastics made of Petro-chemical.

In order to increase the flexibility of polylacticacid, it is possible to add a plasticizer of low molecular weight or integrating polylacticacid with a flexible polymer, for example, poly (butylene adipate-*co*-terephthalate) (PBAT).

Previously, the application of a proper compatibilizer was shown to be significantly crucial for a better variation of properties. Thanks to the accomplishment of a phase morphology categorized with the help of a lower dimension of the dispersed poly (butylene adipate-*co*-terephthalate) (PBAT) phase and an increment in adhesion.

The addition to polylacticacid and poly (butylene succinate) (PBS) was also described for increment in the property of ductility of blends which is based on polylacticacid. Recently, the addition of nano-calcium carbonate to blends made of polylacticacidPLA or polybutylene succinate, permitted for enhancing their qualities maintaining an enhanced influence resistance but integrated with an advanced flexural strength.

Fascinatingly, blends made of polylacticacid or polybutylene succinate with a matrix of polylacticacidPLA were found to describe a partial dimension memory effect that makes them more interesting and fascinating for making films to utilize in secondary packaging.

In addition, tri-continuous blends consisting in polylacticacid (PLA), poly (butylene adipate-*co*-terephthalate) (PBAT), and poly butylene succinate (PBS) (every one of them at thirty-three-point three percentage (33.3%) of weight) were demonstrated to be promising in the process of making bio-based blends having the same properties as poly ethylene.

Never the less, the phase distribution is commonly not easy to handle and makes it stable in ordinary industrial processing. In addition, blends based on polylacticacidare more promising in the context of price and their impact on the environment.

Polylacticacid is completely made of renewable sources. On the other hand, poly (butylene adipate-*co*-terephthalate) (PBAT) and polybutylene succinate (PBS) are partially synthesized at least up to now, from non-renewable resources.

Polybutylene succinate (PBS) will soon be created completely from renewable sources as plants for creating succinic acid from biomass have been enhanced. In the coming time period, also poly (butylene adipate-*co*-

terephthalate) (PBAT) may be renewable because the industrial synthesis of the monomer of poly (butylene adipate-*co*-terephthalate) (PBAT) form renewable resources has been almost fully accomplished.

The main problem in the extrusion is to manage the rate of crystallization, commonly low in polylacticacid, because it impacts the possibility to quickly and effectively cut the extruded strand, therefore easily produced pellets.

In addition, in blown film extrusion, it impacts more of the stability of the process and the achievement of a stiff film. To increase the rate of crystallization, commonly, nucleating agents are added and fascinating effects were seen with the application of both plasticizer as well as a nucleating agent in polylacticacid.

Particularly, it has been seen that the application of oligomeric lactic acid in the form of plasticizer and LAK301 (LAK301 is a derivative of aromatic sulfonate) as a nucleating agent, was much more effective to decrease the time of the whole process of crystallization by half in polylacticacid. And, by using a design of experiment approach, it has been seen that composition range where the half of the time of crystallization was minimized.

In applications like extrusion with the help of flat die or blow extrusion, the fine modification of melt viscosity and melt strength are essential. In these types of cases, the application of chain extenders, much used in the area of polyester processing can be basic or fundamental.

On the other hand, the application of plasticizers with reactive groups having a low number of reactive groups per molecule (two or three) can permit better-modulating polylacticacid-based compounds viscosity, therefore also limiting the migration of the plasticizer from the resulting film thanks to its partial grafting to polylacticacid. The application of a compatibilizer bearing epoxide groups that reduce the melt flow rate of blends made of polylacticacid was also seen to be efficient to enhance the rate of the process by blown film extrusion.

The importance of increment in the melt strength of pure polylacticacid was analyzed by various authors, sometimes with the help of modifiers with epoxide moieties. The process of crystallization of polylacticacid is delayed due to cross-linking because of the decreased mobility of polymer chains.

As the application of chain extenders having a low number of reactive groups per module do not result in extensive cross-linking, the potential of the process of crystallization of polylacticacid is usually not influenced hugely due to a moderate branching also outcomes in an increased rate of the process of crystallization. Integration with an epoxydic extender with a nucleating agent consists of talc and *N*, *N*′-ethylenbis (stearamide).

They demonstrated that a higher kinetic of crystallization can also be accomplished for chain extended and branched polylacticacid produced with a satisfactory quantity of nucleating agents and plasticizers.

Crystallization that took place through the course of process was also observed. Mechanical properties of the optimized films made of polylacticacid achieved by blown extrusion revealed it to be more flexible and ductile as compared to pure polylacticacidPLA. The parameters of blown film extrusion were also seen to impact the biaxially oriented polylacticacidPLA (BOPLA) film properties.

Particularly, it has been noted that when the rate of stretching and the drawn ratio in biaxial-stretching technique is elevated above a particular degree, small δ-crystallites near about the size of ten nanometers with isotropic orientation are created. Therefore there is an enhancement in the toughness of polylacticacidBOPLA film, which is about four times bigger as compared to the film of polylacticacid is created by conventional stretching.

Several papers give proof of the necessity of creating films which are based on polylacticacid with enhanced power of resistance with the help of blown extrusion. However, there is lack of systematic work about a particular enhancement of tear resistance as an outcome of the variation of composition in the context of plasticizer and nucleating agents.

Blends made of polylacticacid are consists of poly (butylene adipate-*co*-terephthalate) (PBAT) or poly butylene succinate (PBS) were used for the production of films with the help of blown film extrusion process.

A reactive plasticizer is applied in the process of extrusion to achieve both, increment in flexibility and to enhance the compatibility. Then the process is scaled up in a semi-industrial extruder to make pellets adding also LAK301 as a nucleating agent.

Films are then made with the help of the process of blown extrusion. The reactive plasticizer and the nucleating agent re applied to accomplish a good command of both, the rate of the process as well as final mechanical properties of the flexible films, with a focus on the tear resistance of films.

7.2 WHAT IS PLA, AND WHAT IS IT UTILIZED FOR?

PLA is unique in relation to most thermoplastic polymers in that it is extracted from inexhaustible resources like corn starch or sugar cane.

Most plastics, on the other hand, are extracted from the refining and polymerization of nonrenewable oil reserves. Plastics that are extracted from biomass (for example PLA) are known as "bioplastics" as has been discussed earlier.

Polylactic Acid is biodegradable and has qualities like polypropylene (PP), polyethylene (PE), or polystyrene (PS). It very well may be delivered from effectively existing equipment of manufacturing (those structured and initially utilized for petrochemical industry plastics).

This makes it generally cost proficient to deliver. In the same manner, PLA has the second biggest manufactured volume of any bioplastic (the most widely recognized is commonly referred to as thermoplastic starch).

There are a huge number of uses for Polylactic Acid. The regular uses include plastic films, bottles, and biodegradable therapeutic equipment (for example screws, pins, bars, and plates that are relied upon to biodegrade inside a year).

PLA contracts under warmth and is in this manner reasonable for use as a therapeutic wrap material. Also, the simplicity with which Polylactic Acid melts takes into account some intriguing applications with regards to 3D printing (particularly "lost PLA casting"). Then again, its low glass change temperature makes numerous kinds of PLA (for instance, plastic containers) which are not meant to hold hot fluid. In light of the regularly increasing measure of plastic wastes worldwide, significant endeavors involving innovative work have been committed to making a solitary use, biodegradable substitute of conventional thermoplastics.

Biodegradable polymers are named a group of polymers that will degrade totally - either into the relating monomers or into items, which are generally part of nature, through metabolic activity of living animals.

They should offer a conceivable option in contrast to conventional, non-degradable polymers as reusing the second one is unfeasible or dangerous to the environment or not affordable. Progress is being made with a polymer called Polylactic Acid (PLA), a moderate, recyclable, creative packaging material produced using sustainable resources.

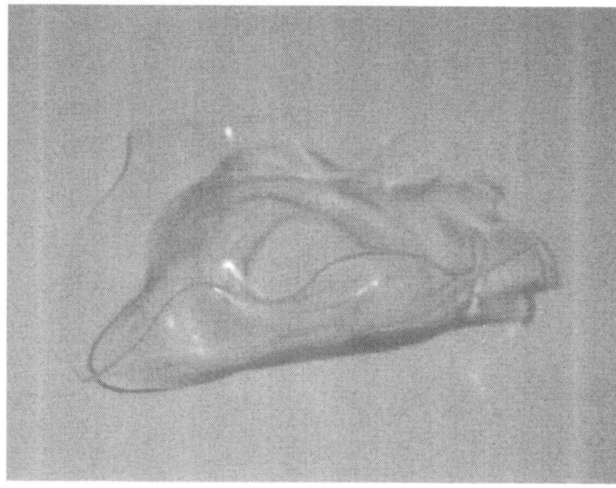

Figure 7.2: PLA cup melted.

Source: https://upload.wikimedia.org/wikipedia/commons/9/94/PLA_cup_melted.jpg

The lactic acid used to make it is extracted from corn. The procedure used to produce it includes both biotechnology and science. PLA has a place with the group of aliphatic polyesters mainly produced using – hydroxyl acids, which incorporate polyglycolic acid or polymandelic acid, and are viewed as biodegradable and compostable.

PLA is a thermoplastic, high-quality, high-modulus polymer that can be produced using yearly inexhaustible resources to yield materials for use in either the modern packaging field or the market of eco-friendly or bio-absorbable therapeutic equipment.

7.2.1 What Are the Different Types of Polylactic Acid and Why is it Used so Often?

There are a few distinct kinds of Polylactic Acid to name such as:

- Racemic PLLA (Poly-L-lactic Acid),
- Regular PLLA (Poly-L-lactic Acid),
- PDLA (Poly-D-lactic Acid), and
- PDLLA (Poly-DL-lactic Acid).

They each have marginally variable qualities. However, they are comparative in that they are created from a sustainable asset (lactic acid: $C_3H_6O_3$) rather than conventional plastics which are extracted from nonrenewable oil.

PLA production is a famous subject as it represents the satisfaction of the fantasy of cost-effective, non-petroleum plastic manufacturing. The biggest advantage of PLA as a bioplastic is its flexibility and the way that it normally degrades when presented to the earth. For instance, a PLA bottle left in the marine ecosystem would ordinarily degrade in the time of six months to two years. Contrasted with regular plastics (which in a similar situation can take a few hundred to a thousand years to degrade) this is really marvelous.

Figure 7.3: PLA in 3D printing cable.

Source: https://www.bb7.com/wp-content/uploads/2017/12/3D_Printing_Featured_2-1-1200x800.jpg

Additionally, there is a high potential for PLA to be exceptionally valuable in short life expectancy applications where biodegradability is exceedingly useful (for example as a plastic water bottle or as a container for foods grown from the ground). Of note, regardless of its capacity to degrade when presented to the components over quite a while, PLA is incredibly powerful in any ordinary application (for example as a plastic equipment part).

7.3 MANUFACTURING OF PLA

Lactide the monomer of polylactic acid is extracted from 100 percent normally sustainable resources. It is for the most part taken out by utilizing plant sugars from field corn as the source material for production.

The monomer of polylactic acid uses under one by twentieth of 1 percent (0.05 percent) of the yearly worldwide corn crop today, and in that capacity contributes almost zero effect on global or local natural food chains. It doesn't require corn; it just needs a sugar source.

Alternate sugar hotspots for lactic acids are Sugar beet, Tapioca, Sugar stick and wheat are 100 percent inexhaustible resources and accessible in great quantity and are cost effective. After this preparation of lactic acid is finished by a fermented broth of sugar and starch utilizing diverse lactic acid bacteria (LAB), exists solely as the dextrorotatory (D or L (+)) structure for manufacturing i.e. polymerization of PLA which is done in three distinct techniques

- Condensation of Lactic acid
- Solid state polymerization and
- Ring-opening polymerization.

Lactic acid (2-hydroxy propionic acid), the single monomer of PLA, is delivered by means of fermentation or chemical treatment. Its two optically dynamic setups, the L (+) and D (−) stereoisomers are delivered by bacterial (homofermentative and heterofermentative) fermentation of sugars.

Industrial lactic acid manufacturing uses the lactic fermentation process instead of synthesis, on the grounds that the engineered processes have many real impediments, including restricted capacity because of the reliance on a by-product of another procedure, not being able to just make the needed L-lactic acid stereoisomer, and high production expenses (Datta and Henry, 2006).

The homofermentative technique is ideally utilized for modern manufacturing since its pathways lead to more prominent yields of lactic acid and to bring down dimensions of by-products.

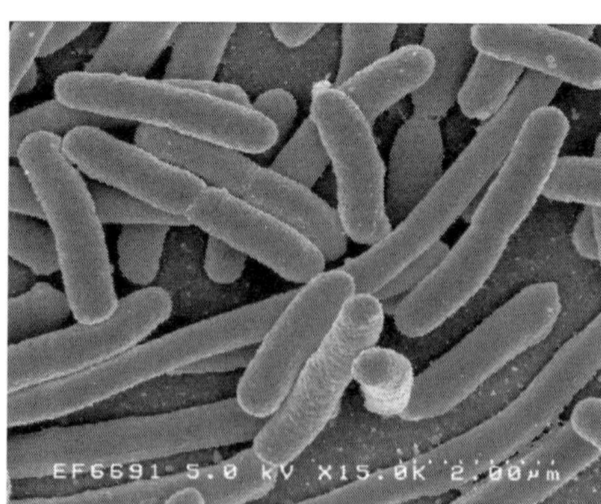

Figure 7.4: Mixed acid fermentation.

Source: https://upload.wikimedia.org/wikipedia/commons/thumb/3/32/EscherichiaColi_NIAID.jpg/913px-EscherichiaColi_NIAID.jpg

The general procedure comprises of utilizing types of the Lactobacillus sort, for example, *Lactobacillus delbrueckii, L. amylophilus, L. bulgaricus*, and *L. leichmanii*, a pH level of 5.4 to 6.4, a temperature span of 38 to 42 °C, and a low oxygen concentration. By and large, unadulterated L-lactic acid is utilized for PLA manufacturing (Mehta et al, 2005).

PLA has a variable sub-atomic weight and its high sub-atomic weight polymer is utilized in the packaging business. Three different ways are workable for the polymerization of lactic acid;

- Direct condensation polymerization;
- Direct polycondensation in an azeotropic compound

An azeotrope is a blend of at least 2 chemical fluids in such a proportion, that its organization can't be changed by simple refining. This happens when an azeotrope is heated up, the subsequent vapor has an indistinguishable proportion of constituents from the mixture that was originally taken; and

- Polymerization through lactide arrangement.

The first strategy depends on esterification of monomers by the help of certain solvents and exudate water is evacuated utilizing dynamic vacuum and high temperatures.

Acquiring high sub-atomic weight polyesters with great mechanical properties, by means of this technique, isn't simple, despite the fact that pre-condensates might be of use for the processing of biodegradable pastes or glues, like the – OH and COOH end groups permit cross-linking with reasonable inorganic or natural multivalent added chemicals (Hartmann 1998).

Delivering high atomic weight PLA polymers by direct polycondensation in an azeotropic arrangement and furthermore, use of certain catalysts is increasingly practicable. The azeotropic arrangement diminishes the refining weights and encourages PLA division from the dissolvable by use of atomic filters.

The diversity and quality of catalysts, dissolvable volume rates, and the response time on the processing of PLA have been contemplated. The outcomes distinguished by utilizing improved equipment for experimentation, the best possible complex catalyst, and dissolvable volume proportion, so as to acquire an atomic weight of PLA of 6.6 × 104 (Li et al, 2006).

Polymerization through lactide development is being achieved in the industries for high sub-atomic weight PLA production. Lactide is a cyclic dimer produced by evacuating water under normal conditions and without dissolvable.

L-lactide, meso (L, D) lactide, and D-lactide are results of L-lactic acid and D-lactic acid. The terms poly lactide and poly (L-lactide) have been utilized in numerous references rather than PLA.

Lactide decontamination is cultivated by vacuum-distillation of high temperatures. After the vacuum-distillation of L-lactide, high sub-atomic weight PLA with a controlled optical and crystal virtue is developed by ring-opening polymerization.

Ring-opening polymerization of lactide can be done in molten liquid or solution by cationic, anionic, and coordination components, contingent upon the initiator used.

The most thought-out dynamic initiator for the L-lactide ring-opening polymerization is stannous octoate (bis 2-ethyl hexanoate, $SnOct_2$), which causes a low level of racemization at increased temperature. It has low harmfulness and is acknowledged by FDA (Puaux et al, 2007).

The decision of initiator framework, co-initiator as chain control agent, catalyst concentration, monomer-to-initiator proportion, and polymerization temperature and time, essentially influence the polymer properties. Properties, for example, atomic weight, level of crystallinity, and leftover monomer part, thus influence the physio-mechanical properties of polylactide and its copolymers.

New thoughts for getting a decline in the products' price of PLA and making production forms all the eco-friendlier, in contrast with prior manufacturing process, incorporate use of leftover from the crops (stems, straw, husks, and leaves) from corn or, possibly, different crops, and utilization of unfermentable deposits as a source for heat, and also substituting some part of power energy by wind energy.

These methodologies decline the utilization of non-renewable energy sources and corn starch as raw materials and furthermore reduce contaminating air, water, and waste outflows to the environment (Vink et al, 2003).

A standout amongst the best purposes of PLA manufacturing in correlation with the other hydrocarbon-based polymers is the lessening of carbon dioxide discharge. Carbon dioxide is accepted to be the most critical factor of worldwide environmental change and its warming. Since carbon dioxide is consumed from the air when corn grows, the utilization of PLA can possibly discharge lesser ozone-depleting chemicals contrasted with competitive hydrocarbon-based polymers. "Net" or "leftover" discharges are determined as absolute outflows from the cradle to the gate of processing plants deducting the absorption of carbon dioxide that happens amid corn production.

This sum is negative for present PLA production. It implies the absolute carbon dioxide utilization from the start of the process to the end of processing is more than its outflow to nature (Bogaert and Coszach, 2000).

Vink and others (2003) came to a conclusion that all created PLA materials go into procedure of composting which transmits carbon dioxide on to the environment; Nevertheless, their net carbon dioxide emanation is lesser than that for various other polymers.

7.4 PLA DEGRADABILITY, BIODEGRADABILITY, AND RECYCLABILITY

Practically all the traditional plastics, for example, PE, PP, PS, and PVC are impervious to microbial damage; despite what might be expected, aliphatic polyesters like PLA are promptly degraded by microorganisms that can be found in the earth.

As indicated by ASTM D6400– 04, a biodegradable plastic is "a plastic that degrades because of the action of naturally occurring microorganisms such as bacteria, fungi, and algae," and a compostable plastic is "a plastic that undergoes degradation by biological processes during composting to yield carbon dioxide, water, inorganic compounds, and biomass at a rate consistent with other known compostable materials and leaves no visually distinguishable or toxic residues."

PLA degradation was examined in animal and human bodies for therapeutic applications like surgical sutures, implants and medication delivery materials (Vainionpaa et al, 1989). In these conditions, PLA is at first degraded by hydrolysis and the dissolvable oligomers developed are processed by cells.

PLA degradation upon transfer in the environment is all the more testing since PLA is to a great extent impervious to damage by microorganisms in soil or sewage under normal conditions. The polymer must first be hydrolyzed at raised temperatures (around 58 °C) to diminish the sub-atomic weight before biodegradation can begin. No degradation was seen on PLA sheets after 6 weeks in soil, indicating that PLA won't degrade in ordinary patio nursery fertilizer (Brandrup et al, 1999; Ohkita and Lee, 2006). Urayama et al (2002) detailed that the atomic weight of PLA films with various optical virtue of the lactate units (100 percent L and 70 percent L) diminished by 20 percent and 75 per cent, individually, after 20 months in soil.

Kale et al (2007) examined the degradation of PLA bottles in a genuine composting condition (fertilizer containing cow compost, wood shavings, and waste material) at 65 °C for 30 days.

They witnessed significant break down, which produces disintegration of the polymer chain into shorter oligomer chains and monomers since the fourth day, and on the 15th day, the containers were at that point in pieces and for the most part comprised of parts from the threaded portions of the caps, and neck (bottle parts having higher thickness) and in the end, on the 30th day, the bottles were found to be totally degraded. Microbial and enzymatic degradation of PLA has as of late been concentrated by numerous analysts on the grounds that these kinds of degradations generally need not bother with the high temperatures, to be achieved.

Williams (1981) first detailed the degradation of PLLA by proteinase K from *Tritirachium album.* A short time later numerous examinations were accomplished for finding diverse chemicals relating to PLA degradation.

Detailed chemicals that empower to degrade PLA in various scale incorporate, alkaline protease (Oda and others 2000), serine proteases, for example, subtilisin, trypsin, elastase, and α-chymotrypsin (Lim and others 2005), Cutinase-like compound (Masaki et al ,2005).

Lipase could hydrolyze low atomic weight PLLA and a few copolymers, for example, PDLLA (poly D, L-lactic acid) and, poly(D-lactid-co-glycolide) but not PDLA (poly D-lactic acid) and high sub-atomic weight PLLA (Fukuzaki et al ,1989).

Pranamuda et al (2001) found a protein from Amycolatopsis sp. groups and named it PLLA depolymerase. The ideal pH and temperature for this chemical were 6.0 and 37 to 45 °C, separately. PLLA depolymerase can additionally hydrolyze casein, silk fibroin, succinyl-p-nitroanilide, but not PHB and PCL.

The enzymatic degradation of aliphatic polyesters by hydrolysis is a procedure that is carried out in two steps. The first step involves adsorption of the enzyme on the outside of the substrate through surface-binding and the second step involves hydrolysis of the ester bond (Tokiwa and Calabia, 2006).

Pranamuda et al (2001) first detached a PLA-degrading microorganism of Amycolatopsis strain from soil condition, which was equipped for degrading 60 per cent of the PLA film after 14 days.

Suyama et al (1998) revealed that PLA-degrading microorganisms are not dispersed in the natural habitat in huge quantities and, in this way, PLA is less vulnerable to microbial damage in the natural habitat than other engineered aliphatic polyesters like PHB, PCL, and Poly (butylenes succinate) (PBS).

7.4.1 PLA Market size and Forecast

Restricted Fossil Fuel Resources, rising petrochemical costs, Reinforcement of PLA with different materials, ecological health awareness and motivating activities by the government will be key elements driving the PLA market.

Fortification of PLA with different materials makes tremendous opportunities for the future. As natural fibers are very cheap and affordable and degradable, it's supported with PLA prompts a material which is sustainable and also environment-friendly.

The worldwide PLA market is estimated to achieve a market size of 5.2 Billion USD, developing at a CAGR of 21.6 percent amid 2013 to 2020. By 2020, the share of the overall industry of PLA is anticipated to rise to 25 to 30 percent.

7.4.2 Cost of PLA

In the worldwide market, the production capacity of PLA and the market demand rises but its cost is on the decrease. Cost of PLA has decreased from the initial value of USD 1.50 per lb. in the mid-2000's to as of now USD 0.85 per lb. or lower, making PLA competitive with materials, for example, PET which cost about USD 0.65 per lb.

The expense contrasted with normal plastic items is ending up being increasingly competitive as the interest for biodegradable item rises. Additionally, expenses of PLA may not vacillate as much as oil-based resins as it is produced using corn rather than petrochemicals and subsequently protected from unstable and rising oil costs.

By and large, for a properly evolved Polylactic acid industry and to additionally cut down the cost, the components, for example, extending the manufacturing scale, improving the properties of PLA and backing from the administration will turn out to be essential.

7.5 ADVANTAGES AND DISADVANTAGES OF PLA

7.5.1 Advantages

- One of the significant points of interest of PLA is its biodegradable nature and ease of preparation and handling.
- Under the correct conditions, PLA can be broken down in under a month rather than the hundreds of years it will take for conventional plastics to deteriorate. PLA is particularly appropriate in short life expectancy applications, for example, in water bottles and food containers.
- The procedure by which PLA is made is more environmentally friendly. Notwithstanding utilizing inexhaustible raw materials, the release of greenhouse gases amid manufacturing is additionally lower.

Figure 7.5: A shopping bag made of PLA blended bio flex.

Source: https://upload.wikimedia.org/wikipedia/commons/0/08/Shoping_Bag_made_of_PLA-Blend_Bio-Flex.jpg

- Since carbon dioxide is consumed amid the development of corn, the net greenhouse gas discharge of the general PLA manufacturing procedure can even be viewed as negative. Progressing research about the utilization of alternate sugar sources, for example, farming and family unit wastes, even recommend that PLA production can prompt a lessening in overall solid waste.

- PLA is a thermoplastic, which means it will transform into a fluid in its softening purpose of 150 to 160 Celsius. A unique component of thermoplastics is that they can be warmed, set after cooling, and warmed again to develop different shapes with no degradation.
- Conversely, a thermosetting plastic, (for example, epoxy or melamine) must be warmed and formed once, however, the subsequent item can never again be warmed as it will simply burn. This property of PLA makes it an alluring material for reuse.
- PLA can be separated to its unique monomers by a warm de-polymerization process or by hydrolysis. The subsequent monomer solution can be cleaned and utilized for ensuing PLA production with no loss of value. Should a material produced using PLA be burned, no dangerous vapor will be created.

7.5.2 Disadvantages

- The simplicity with which PLA softens makes it a material that is anything but difficult to work with. In any case, this additionally renders PLA unacceptable for high-temperature applications, for example, containers made for storing hot fluids. A material produced using PLA may even hint at getting delicate or twisting on a sweltering summer day.
- PLA is still viewed as mediocre compared to polyethylene terephthalate (PET) for long term food stockpiling applications because of permeability issues with PLA.
- PLA packaging material has been observed to be progressively penetrable to dampness and oxygen as compared to different plastics, which may result in quicker food deterioration. Applications, where toughness and effect opposition are crucial, may additionally not be good for the weaker PLA.

7.6 WHAT MATERIALS CAN BE MADE WITH PLA?

PLA can be prepared by means of extrusion, injection molding, casting, blown film, thermoforming, and fiber turning to develop valuable and flexible items. It is regularly accessible in the market as thin films for thermoforming, plastic pellets for injection molding, or 3D printable fibers. A diversity in colors of PLA can be found.

Tougher forms of PLA have been created by blending diverse PLA isomers, bringing about higher liquefying temperature (higher by 40 to 50 Celsius) and increased mechanical quality. This improved form has seen a wide cluster of uses, for example, microwavable containers and building plastics.

Figure 7.6: Pinatex, the fibers of pineapple hanging on rope to be dried.

Source: https://upload.wikimedia.org/wikipedia/commons/7/7f/After_the_fibre_from_the_pineapple_leaves_is_decorticated_and_washed_the_farmers_hang_it_up_to_dry.jpg

- Transient applications, for example, food containers, water bottles, and expendable cutlery are a well-known use for PLA. A PLA film decreases in size after warming, making it an alluring material for shrink wrap.
- Comments from firms that utilize PLA as material for food packaging demonstrates the inclination for PLA because of increased feel, better printability, great protection from grease

and oils, and diminished issues in taste and smell transfer.

- PLA is broadly utilized in the restorative field because of its capacity to degrade into non-harmful lactic acid. Therapeutic implants, for example, screws, bars, sticks and work have been made utilizing PLA. Inside the patient's body, these inserts totally separate in a half year to 2 years, dispensing with the requirement for a further medical procedure.
- PLA can be taken out into dainty filaments with noteworthy mechanical quality. These PLA strands have been utilized to fabricate easygoing games attire, upholstery material, cleanliness items, and diapers.
- PLA is one of the two plastics most ordinarily utilized in 3D printing (the other one being Acrylonitrile Butadiene Styrene or ABS). In particular, PLA is generally utilized in intertwined fiber production 3D printing, where PLA solids are encased in mortar like moldings to shape forms that can be weighted up with liquid metal. This is a strategy known as "lost PLA casting".

7.6.1 Crystallization of Bioplastics

Biodegradable poly (lactic acid) (PLA) more often than not has crystallinity under 10 percent because of its poor crystallization capacity. In this work, we discovered high weight could essentially encourage the development of crystallites of PLA, bringing about a crystallinity high up to 66.3 percent at pressure and temperature of 300 MPa and 185 °C.

High-weight incited crystalline redesign and lamellar thickening prompted two liquefying temperatures in the profoundly solidified PLA but without cold crystallization contrasted with the typical weight solidified PLA. Temperature reliance of high-weight crystallization of PLA recommended desirable crystallization temperatures for exceedingly solidified PLA items.

Biodegradable poly (lactic acid) (PLA) has been broadly utilized in business and biomedical applications for its great processability and inexhaustibility. The properties of PLA items are altogether influenced by its crystallinity and crystalline morphology.

It was accounted for that the heat damaging temperature of PLA with 40 per cent crystallinity is around 15 °C higher than that of amorphous PLA. However, for its natural low crystallization capacity, PLA is regularly in a shapeless state after normal processing, for example, extrusion or injection molding. In light of low crystallinity just as low glass progress temperature, PLA indicates exceptionally poor heat resistance, which seriously confines its functional applications. Hence, the expansion of PLA crystallinity turns into the bottleneck issue that impedes the utilizations of PLA items.

The crystallization of polymers at high weight has pulled in incredible interests since it can deliver very much characterized crystals with high liquefying point and crystallinity, subsequently viably improving the mechanical properties just as the thermal behavior.

For instance, the crystallinity of high-weight solidified polbuthylene can be up to 90 percent, particularly better than the normal handling circumstances. The dynamic Young's modulus of weight altered polyethylene is twice as high and its quality is greater.

Normally, one could expect that the high weight processing technique can be used to improve the crystallinity of PLA.

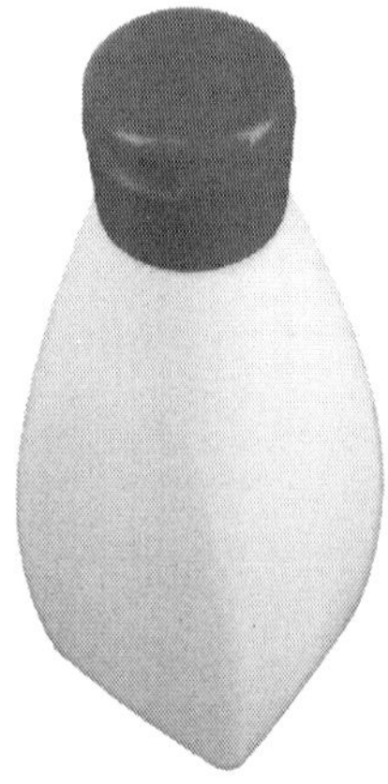

Figure 7.7: Shampoo bottle made of PLA material.

Source: https://upload.wikimedia.org/wikipedia/commons/c/c4/Shampoo_Bottle_made_of_PLA-Blend_Bio-Flex.jpg

7.6.2 Crystal Structure and Single Crystals

Crystal Structure and Single Crystals L-lactic acid is a monomer that can be acquired from common sources. It merits calling attention to that (S, S)-lactide is alluded to as L-lactide (PLLA), (R, R)- lactide relates to D-lactide (PDLA) and (S, R)- lactide is a meso-lactide (PLDA).

Distinctive sorts of PLA crystals can be acquired relying upon the crystallization conditions. The a-structure is the crystal type most commonly found for PLA. It was first detailed by De Santis and others. They verified that the a-structure crystal compares to a helical adaptation by X-beam fiber photos.

The a-structure crystal comprises of 3.3 monomers per turn and a monomer rehash unit of 27.8 nm. Zhang and others revealed another change called a0 - structure crystal, a structure like the a-structure however with packing that is lesser dense.

The a0 - structure crystals were extracted at crystallization temperatures (Tc) beneath 120 °C. Different investigations propose that a0 - structure crystals structure underneath 100 °C, while in the temperature scope of 100–120 °C both a and a0 - structures can emerge.

Eling and others observed b-structure crystals, whose dissolving temperature is 10 °C lower than the softening temperature of the relating a-structure. Cartier and others revealed g-structure crystals, which are acquired by epitaxial crystallization of PLA in hexamethylbenzene.

In g-structure, two chains are antiparallel to the crystal unit cell. Fischer and others arranged lactide copolymers (meso-lactide) single crystals containing 2.75 percent and 10.2 percent D-units. The PLDA single gems show a progressively standard terraced morphology for lower D-unit content.

However, they were not shaped when the chemical of D-units was more prominent than 15 percent. Kalb and others arranged PLLA single gems developed from weakening solutions (0.08 percent by weight). Lamellae were very much characterized capsule molded at Tc ¼ 55 1C, with thicknesses around 10 nm.

At shallower undercooling, for example, Tc ¼ 90 1C, lamellae were truncated capsules with thicknesses around 12 nm.

The PLLA tests were solidified from the liquefy at 164 °C, and the single gems were hexagonal-molded. An equimolar blend of PLLA/PDLA was first concentrated by Okihara (1991) et al.

Their outcomes demonstrated that PLLA/PDLA can co-solidify into a stereo-complex, whose crystal framework is triclinic. Besides, et al (1997), proposed a structure wherein the chains of PDLA and PLLA structure 31 and 32 helices, separately.

In view of the stereoregularity (L and D) of lactic acid, L, D and D-L type lactides can be utilized in PLA production. Semi-crystalline PLLA (poly-L-lactic acid) or nebulous PDLA (poly-D-lactic acid) can be created by a ring opening polymerization of these lactides.

For business, PLA, a mix of a higher measure of L-lactide and a lower measure of D-lactide is utilized. The proportion of the two lactides impacts the mechanical properties, crystallinity and even the degradation qualities.

At the point when semi-crystalline PLA is prepared, its crystallinity changes rely upon the cooling rate. Ordinarily in injection forming, because of the high cooling rates, indistinct PLA items are delivered.

In the differential filtering calorimetry (DSC) bend of this shapeless PLA an exotherm can be discovered, identified with the cool crystallization above Tg, and an endotherm, identified with dissolving. At the point when PLA is recrystallized before estimation, the exotherm in the DSC bend is absent.

Amid the crystallization of PLLA at high temperatures (>130– 140°C) it is discovered that the disarranged α' crystalline structure, which can be created by toughening at lower temperatures, changes into the arranged α structure.

The contrasts between the α' and α structures can be seen by wide edge X-beam diffraction (WAXD) spectrometry and DSC examination. In the WAXD investigation, increasingly extreme and new pinnacles can be found, while in the DSC bends two endothermic pinnacles seem to identify with the liquefying of the α'-and α stages.

The crest at the lower temperature is identified with the softening of the α' structure and its recrystallization into the α structure, while the second pinnacle relates to the dissolving of the α structure. The two procedures, the softening of the α' structure and the recrystallization into the α structure can be considered as the α'– α stage progress.

For lower sub-atomic weight PLA, the α'– α stage change happens at lower temperatures contrasted with higher sub-atomic weight PLA. Crystallinity additionally impacts the dynamic mechanical investigation (DMA) properties.

As it very well may be found in the DSC bends, the recrystallization happens above Tg and builds the capacity modulus. Recrystallized PLA items additionally convey a gigantic preferred standpoint contrasted with shapeless items.

This preferred standpoint is shown in the lot higher Heat direction temperature (HDT); while shapeless PLA items have an HDT of 50°C;

semi-crystalline PLA items have an HDT of 130°C. In spite of the fact that PLA can be recrystallized in the wake of handling into an item, the items shrivel, and noteworthy warpage happens amid recrystallization because of the increasing crystallinity.

Contortion free semi crystallized PLA items must be made amid the procedure utilizing nucleating specialists. The powder was observed to be a viable nucleating specialist, which was seen amid DMA examinations.

The injection molding evaluation PLA with a D-Lactide chemical of 4 percent can be obtained in semi-crystalline structure, however when it is handled into an item, or blended with regular fillers or strands utilizing melt blending systems like extrusion, this PLA winds up shapeless even at low cooling rates (without the utilization of nucleating specialists).

Accordingly, in the event that it is to be additionally prepared like on account of PLA based composite pellets or on the off chance that it is to be reused like in a mechanical handling plant where the rejected parts are to be reused, at that point in light of the adjustment in crystallinity, a very surprising material will be prepared.

The adjustments in crystallinity amid handling and its impact on the re-processing have been explored with injection molding; moreover, the properties of different crystalline chemical injection forming grade PLA gums were dissected, and the handling troubles because of the crystalline structure were clarified.

7.7 OVERALL CRYSTALLIZATION KINETICS

From isothermal crystallization tests performed by DSC, the opposite of the test half-crystallization time can be acquired, an amount that is relative to the general crystallization rate since it incorporates commitments from essential nucleation and development. This demonstrates the backward of half-crystallization time (t50 percent) as a component of crystallization temperature for various PLA tests (diverse atomic weights and D-isomer concentration).

A portion of the bends displayed additionally contain the two maxima that have a similar beginning as in. Di Lorenzo (2007), demonstrates that PLAs show intermittence in crystallization rate at around 118 1C. For the most part, the opposite of half-crystallization time diminishes with the D-isomer content, as detailed by Kolstad (1996), who demonstrated that the decrease is roughly 45 percent with increasing 1 percent in meso-lactide content. Then again, the impact of the sub-atomic weight and comparable D-isomer focus demonstrates that the opposite of half-crystallization time diminished with sub-atomic weight as a result of the low chain versatility.

Despite the brokenness revealed in certain bends, it may compare with the double pinnacles detailed for the spherulite development rate plots, however, the outcomes are not as obvious. This is most likely the aftereffect of the impact of nucleation on the general crystallization rates.

Auliawan et al (2012), examined ternary polymer mixes of PLLA, poly (methyl methacrylate) (PMMA) and poly (ethylene oxide) (PEO) as a lattice for nanocomposites. Their decisions were that Nano clays upgraded the non-isothermal crystallization of the mix since this ternary polymer mix scarcely solidified when cooled from the dissolve.

The expansion of vermiculite at weightings bigger than 1 percent by weight hindered the crystallization procedure, while montmorillonite upgraded the crystallization procedure to some degree before impeding it.

Mixes of PLA with a dendritic hyperbranched polymer (DHP) and starch were considered by Zhang and others. They demonstrated that expansion of DHP and starch increased both the cool crystallization capacity and the isothermal crystallization rate of PLA.

Powder and triphenyl phosphate (TPP) have been accounted for as nucleation specialists for PLLA. The isothermal energy of flawless PLA and its mixes with TPP and powder were accounted for by Xiao et al (2009), . Nonetheless, the crystallization rate diminished with the consolidation of TPP.

Furthermore, Quero and others examined the isothermal cold-crystallization of PLA/PBAT mixtures with and without acetyl tributyl citrate (ATBC). SEM results demonstrated that the mixtures displayed two stages, but halfway miscibility: changes in both, Tg and Tm, of the PLA stage, a relationship of the spherulitic development rate with the mix solution and the impediment of PBAT beads inside PLA spherulites.

ATBC acted like a plasticizer for the two stages (slick PLA and PBAT). On account of mixes, ATBC wants to be incorporated inside the PBAT-rich stage. There was a synergistic impact on the general crystallization rate of PLA when both ATBC and PBAT were available in the mixture.

7.8 MATERIALS, PROCESSING AND EXPERIMENTAL

Semicrystalline Polylacticacid injection molding grade resin (type Ingeo 3051D from Nature Works Ltd., USA, with a D-lactide content of four percent) was applied to perform the measurements. Semi crystalline Polylacticacid was dried at a temperature of eighty-five degrees celsius for six hours before injection molding to avoid any deprivation because of the hydrolysis.

Two-millimeter-thick, eighty millimeters by specimens were injection molded with the help of an Arburg Allrounder 320C 600-250 injection molding machine armed with a diameter of the thirty-five-millimeter screw.

The injection pressure was eight hundred and fifty bars, on the other hand the applied holding pressure was six hundred bars. The temperature profile used was 165-175-180-185-190°C from the hopper to the nozzle. To make quality Polylacticacid products some precautions were taken. To form a homogenous melt, at least thirty bars of backpressure was necessary with a screw rotational speed of fifty meters per minute.

In some cases, when the mold opened, the sprue broke, stuck into the stationary mold half and had to be removed manually. To prevent this, a long holding time of twenty seconds had to be used, not only to compensate for the shrinkage of the part but also for the shrinkage of the sprue, specifically to prevent the creation of cavities.

In addition, the part and the sprue had to be sufficiently stiff to prevent this process, therefore for proper de-molding, long holding and residual cooling times of thirty seconds and a low mold temperature of twenty-degree celsius had to be used.

The injection molded samples were totally transparent, indicating their amorphous nature. After injecting the samples, fifty-millimeter-long and two millimeters by four-millimeter specimens were cut for the Dynamical Mechanical Analysis measurements, and two-millimeter-thick fifteen millimeters by fifteen-millimeter specimens were cut for the wide-angle X-ray diffraction (WAXD) measurements.

The other injection molded samples were shredded, repalletized, and the amorphous Polylacticacid pellets were examined by the implementation of two different approaches. Firstly, it was processed again in the same way as before to analyze the differences in processing. Secondly, the pellets were strengthened at a temperature of eighty-degree celsius for only ten minutes to detect the beginning of the process of re-crystallization with a scanning electron microscope (SEM) type Jeol JSM6380 LA.

The specimens (for wide-angle X-ray diffraction (WAXD), Dynamical Mechanical Analysis (DMA) and Differential scanning calorimetry (DSC)) were placed on a glass plate in a vented oven for annealing at 60-80-100-120-140°C for 10-20-30-40-50- 60 minutes.

The reason for the glass plate support was the low heat deflection temperature of the material. The notation of the samples is as follows: X/Y, where X represents the temperature [°C], and Y represents the interval of time [min] of the strengthening.

The Dynamical Mechanical Analysis (DMA) measurements were done with the help of a PerkinElmer Pyris Diamond dynamic mechanical analyzer, whereas the Differential scanning calorimetry (DSC) measurements were made with the help of a Perkin-Elmer Pyris 1 differential scanning calorimeter.

The crystallinity was evaluated from the Differential scanning calorimetry (DSC) melting enthalpies, through the application of a melting enthalpy value of seventy-five-point-five-seven Jules per gram (75.57 J/g) for hundred percent crystalline poly-L-lactide (PLLA).

The un-annealed specimens were measured at heating rates of 1-2-5-10-15-20°C per min, whereas the annealed specimens were measured at a heating rate of five-degree celsius per minute. wide-angle X-ray diffraction (WAXD) analysis was performed with the help of a Philips model PW 3710 based PW 1050 Bragg-Brentano Para focusing goniometer using CuKα radiation ($\lambda = 0.15418$ nm).

In addition, the analysis of the crystalline structure of pure Polylacticacid, thirty weight percentage maize starch (type Meritena 100, from Brenntag Hungária Ltd., Hungary) content Polylacticacid was produced with the help of a Brabender PlastiCorder PL 2100 twin screw extruder. The processing of the annealed and un-annealed blends of Polylacticacid or starch was analyzed.

7.9 CONCLUSION

In the world that focuses on the business aspects of everything, the concerned people find it important to bring a reduction in the footprint that concerns the environment. This indicates the increase in sustainable use of products and resources along with decreasing the costs of the products.

It is a good thing that the possibility of accomplishing both the targets is always there, owing to the growing number of uses of Polylactic Acid. It is increasingly being employed in the packaging industry and the latest developments in the industry prove the conclusions.

There are a lot of benefits that the use of PolylacticacidPLA has in the industry catering to the packaging of products and food. There are some advantages that PLA has over the products derived from Petrochemicals.

In comparison, PLA is biodegradable since it is taken from the sources that are thoroughly natural. It has lesser emissions in terms of greenhouse gases and leaves a very reduced carbon footprint. It is also very simple to manipulate the relationship between structure, property and processing in the PLA. There are some more advantages that may exist.

Moving on to the packaging materials that involve the use of PLA will gain more and more significance in the event that the governments, various different industrial and business houses and the general people, grow their awareness in terms of being protective and secure towards the environment and they formulate policies that are environmentally friendly.

All these things coupled with the decreased costs make the PLA a good source of plastic products. In the current scenario and looking forward towards the future, the significance of PLA in the packaging industry is most likely to rise in stature, which will ultimately replace the products made out of polystyrene and other such conventional materials that have been traditionally been used in the packaging industry.

REFERENCES

1. 360 Packaging. (2018). *Polylactic Acid: A Sustainable Bioplastics Packaging Option.* [online] Available at: https://packaging360.in/insights/polylactic-acid---a-sustainable-bioplastics-packaging-option [Accessed 29 Mar. 2019].
2. Flynt, J. (2017). *What is PLA? Polylactic acid properties, uses, & melting point - 3D Insider.* [online] 3D Insider. Available at: https://3dinsider.com/what-is-pla/[Accessed 29 Mar. 2019].
3. Jamshidian, M., Tehrany, E., Imran, M., Jacquot, M. and Desobry, S. (2010). Poly-Lactic Acid: Production, Applications, Nanocomposites, and Release Studies. *Comprehensive Reviews in Food Science and Food Safety*, [online] 9(5), pp.552-571. Available at: https://onlinelibrary.wiley.com/doi/full/10.1111/j.1541-4337.2010.00126.x [Accessed 29 Mar. 2019].
4. Mallegni, N., Phuong, T., Coltelli, M., Cinelli, P. and Lazzeri, A. (2018). Poly (lactic acid) (PLA) Based Tear Resistant and Biodegradable Flexible Films by Blown Film Extrusion. *Materials*, [online] 11(1), p.148. Available at: https://www.ncbi.nlm.nih.gov/pmc/articles/PMC5793646/ [Accessed 29 Mar. 2019].
5. Müller, A., Ávila, M., Saenz, G. and Salazar, J. (n.d.). CHAPTER 3. Crystallization of PLA-based Materials. *Polymer Chemistry Series*, [online] pp.66-98. Available at: https://www.researchgate.net/publication/268578390_Crystallization_of_PLA-based_Materials [Accessed 29 Mar. 2019].
6. Rogers, T. (2015). *Everything You Need to Know About Polylactic Acid (PLA).* [online] Creativemechanisms.com. Available at: https://www.creativemechanisms.com/blog/learn-about-polylactic-acid-pla-prototypes [Accessed 29 Mar. 2019].
7. Tabi, T., Sajo, I., Szabo, F., Luyt, A. and Kovacs, J. (2010). Crystalline structure of annealed polylactic acid and its relation to processing. *Express Polymer Letters*, [online] 4(10), pp.659-668. Available at: http://www.expresspolymlett.com/articles/EPL-0001596_article.pdf [Accessed 29 Mar. 2019].
8. Zhang, J., Yan, D., Xu, J., Huang, H., Lei, J. and Li, Z. (2012). Highly crystallized poly (lactic acid) under high pressure. *AIP Advances*, [online] 2(4), p.042159. Available at: https://aip.scitation.org/doi/full/10.1063/1.4769351 [Accessed 29 Mar. 2019].

Chapter 8

SUSTAINABILITY OF BIO MATERIALS

CONTENTS

Worries over non-sustainable fossil fuel supply and environmental change have been driving the *Renaissance* of bio-based materials. To substantiate natural cases, the effects of bio-based materials are regularly measured by applying life cycle assessment (LCA). The universally concurred LCA standards gives conventional suggestions on the most proficient methods to assess the natural effects of items and administrations however don't address subtleties that are explicitly important for the existence cycles of bio-based materials. This chapter provides an overview of key issues and methodologies explicitly pertinent to the LCA of bio-based materials. Currently, used plastics are mostly produced from petrochemical products, but there is a growing demand for eco-friendly plastics. The utilization of bio-based plastics, which are created from inexhaustible assets and biodegradable plastics, which are degraded in the earth, will prompt an increasingly economical society and help us fathom worldwide ecological and squander the board issues.

8.1 INTRODUCTION

The present direct 'take, make and arrange' financial model depends on expansive amounts of modest, effectively open assets and vitality, and is achieving its natural breaking points. In a roundabout economy, assets are recovered in the natural cycle and recuperated and reestablished in the specialized cycle. This is an alluring and feasible option monetary model that organizations have just begun investigating today. Fundamental drivers for change are societal difficulties like environmental change, asset limit, vitality instability, squander issues and an expanding population with developing product demands.

Bio-based materials (for example materials produced using inexhaustible natural assets) hold extraordinary guarantee to wind up maintainable options for fossil-based materials because of sustainable feedstock use, the probability to utilize (enzymatic) low vitality generation courses and conceivable elective waste systems, similar to biodegradation.

Additionally, bio-based materials are not characteristically supportable. The biomass feedstock supply, fabricating procedure, interdependency and rivalry with other item esteem chains, and waste situations assume a vital job in the dimension of maintainability. Also, manageability issues like the challenge of biofuels creation with their sustenance security have just been distinguished and ought to be dealt with when further building up the bio-based economy. The progress to a bio-based economy is driven by the general population and private interest for progressively feasible items and by legislative bio-economy arrangements. In addition, up and coming around economy arrangements and the Paris atmosphere gathering (COP21) assertion further lift the advancement of a manageable bio-based economy.

Be that as it may, the bio-based economy is still in the beginning period of advancement. Mechanical and advancement challenges must be handled to create practical and focused bio-based items, along these lines to make the bio-based economy a reality. Principle challenges are to create suitable new esteem chains and items dependent on inexhaustible feedstocks, to create imaginative bio-based items for distinguished market applications and to make and quicken the market take-up of bio-based items and applications.

Plastic materials are, at present, considered imperative materials because of their outstanding properties and execution over different materials, for example, metal and wood. It is anticipated that the interest in plastics will keep on rising after a pattern that has expanded since the 1950s. As plastic materials are of such significance in our general public and realizing that materials are an essential determinant of maintainability, the substitution of oil-based plastics with bio-based plastics is viewed as a promising option since it will decrease the reliance of plastics on petroleum products and the weight on landfills from plastic waste.

Advancement and commercialization of bio-based plastics for an assortment of employment in items and packing is likewise of incredible enthusiasm as makers are searching for more secure and more advantageous materials as substitutes for synthetic concoctions and materials of concern in buyer items. Bio-based materials are promising as their feedstocks are inexhaustible, hypothetically, they can be treated; the soil or reused, and their generation procedure might be more vitality productive than oil-based plastics handling.

This chapter gives understanding of the wellbeing and natural effects of bio-based plastics. Economical materials are those that diminish effects to word related and general wellbeing just as to nature for an amazing duration cycle. In this way, the point of this examination was to subjectively assess the general manageability of various bio-put together plastics based with respect to a broad survey of the writing and thinking about ecological, wellbeing and security impacts for an amazing duration cycle. A figure called the Bioplastics Spectrum was made to outwardly condense the information and to aid basic leadership about bio-based plastics determination.

8.2 DEFINING A SUSTAINABLE BIO-BASED PLASTIC

An agreement definition for a bio-based plastic does not yet exist. The Business-NGO Working Group for Safer Chemicals and Sustainable Materials (an extraordinary joint effort of business and NGO pioneers who advance the utilization of more secure synthetics in purchaser items) characterizes bio-based plastics as "plastics in which 100% of the carbon is gotten from inexhaustible farming and ranger service assets, for example, corn starch, soybean protein and cellulose" (Business-NGO Working Group for Safer Chemicals and Sustainable Materials, 2007).

The US Department of Agriculture characterizes bio-based plastics as "business or modern merchandise, (other than feed or nourishment), created in entire or in huge piece of natural items, ranger service material, or sustainable household farming materials, including plant, creature or marine materials". ASTM characterizes a bio-based material as "a natural material in which carbon is gotten from an inexhaustible asset through organic procedures" (The Biodegradable Products Institute, 2006).

Bio-based materials incorporate all plant and creature mass got from CO_2 as of late fixed by means of photosynthesis, per the meaning of a sustainable asset. In all actuality, most bio-put together plastics with respect to the market today are mixes of bio-based and oil-based materials. A bio-based plastic isn't really an economical plastic; this relies upon an assortment of issues, including the source material, creation procedure, and how the material is overseen toward the finish of its helpful life.

Because of the myriad issues to consider, it can be very challenging to determine which plastic materials are safest and healthiest for workers, consumer, and the environment. Over the last decade, researchers and citizen advocates have developed several tools to assist in decision making about plastics selection. The plastics pyramid (Fig. 8.1) developed by Thorpe and Van der Naalde in 1998 was an early attempt to visually display the life cycle hazards of different plastics to assist in materials selection.

This ranking focused on the toxicity of the material, considering production hazards, use of harmful additives, hazards in use, and disposal hazards. In this pyramid, bio-based polymers form the bottom of the pyramid, indicating they are most preferable, as they are made from renewable resources, and theoretically are biodegradable and compostable. The Plastics Spectrum, created in 2005 (Rossi et al., 2005), added to the information presented in the Plastics Pyramid by displaying the recyclability of plastics alongside their life cycle hazards (Fig. 8.2).

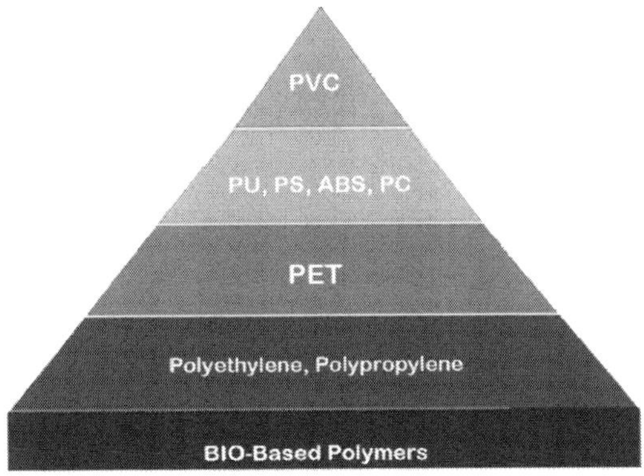

Figure 8.1: The plastics pyramid.

Source: Rossi et al., 2005

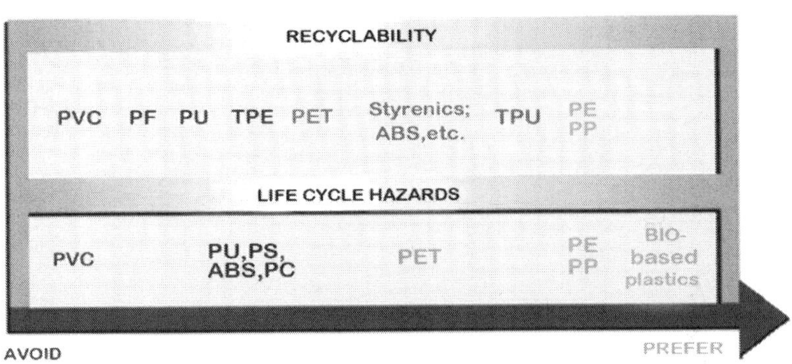

Figure 8.2: The plastics spectrum.

Source: https://www.researchgate.net/figure/The-plastics-spectrum-Source-Rossi-et-al-2005_fig2_229313085

Since the Plastics Pyramid was created, bio-based plastics are a lot further along in their business advancement. These materials should be assessed cautiously for supportability. Rossi and Lent (2006) have built up an Environmental Preference Spectrum for the medicinal services industry. This range is the first to take the class of "bio-based" and start to give criteria to what makes these materials ideal. This range inclines towards bio-based plastics that are reasonably developed that can be promptly reused in view of the existing foundation. It recommends dodging plastics that are fabricated with, contain, or radiate profoundly perilous synthetic substances and keeping away from those that are not promptly recyclable. This report characterizes "reasonably created bio-based materials as those that seem to be developed without hereditarily altered living beings (GMOs), unsafe pesticides, affirmed as supportable for the soil and biological systems, and compostable into sound and safe supplements for sustenance crops".

All the more, as of late Rossi and Greiner (2009) built up the Plastics Scorecard to rate distinctive sort of plastics dependent on their life cycle effects and perils to human wellbeing and condition. The three center components include economical feedstocks, green science and shut circle frameworks.

In this score card, plastics get an evaluation from F (poor life cycle execution) to Aþ (great life cycle execution). For instance, the greatest achievable evaluation for PLA is A when corn is developed without GMOs and atrazine pesticide. PLA's evaluation is C when GMOs and atrazine are utilized. Interestingly, the most extreme achievable evaluation for PVC is F considering its synthetic discharges and breakdown items (dioxins and furans that are tireless, bioaccumulative, dangerous and cancer-causing synthetic compounds).

8.3 PRINCIPLES FOR SUSTAINABLE BIOMATERIALS

The Sustainable Biomaterials Collaborative (SBC, 2009) is a system of associations that were framed to goad the improvement and utilization of biomaterials that are reasonable for an incredible duration cycle. Standards for the improvement of manageable biomaterials, as created by this association, include:

- Diminish the measure of material, item and packing utilized.
- Dispense with single-use items that can be neither reused not treated with soil.
- Stay away from non-renewable energy source-based materials for materials and items got from sustainable feedstocks.
- Address manageability over the existence cycle of the material: the developing of the feedstock, assembling of the biomaterial and last item, utilizing the item and recovering the material toward the finish of its unique use.

- Characterize supportability to incorporate issues of condition, wellbeing, and social and monetary equity.
- Structure and use items that are reusable, recyclable or compostable
- Empower agrarian frameworks that are reasonable for rancher's nature, farmworkers and networks. This way to take out perils of concern amid feedstock creation, moderate, ensure and fabricate soil, save supplement cycles, secure air and water access and quality, advanced natural decent variety, decrease in general vitality use and its effects, diminish transportation impacts, create and affirm a far-reaching economical agribusiness plan, secure specialist's wellbeing and give reasonable remuneration.
- Bolster little far sized family possessed and worked ranches.
- Try not to utilize GMOs in rural feedstock creation.
- Use synthetic concoctions that meet the 12 Principles of Green Chemistry. These standards seek after the structure and creation of synthetic substances in a way that limits the potential for wellbeing and natural dangers.
- Dodge built nanomaterials and synthetic substances that have not been tried for natural and general wellbeing impacts over the existence cycle; and
- Decentralize creation and purchase locality to lessen the ecological impression of generation, transportation and utilization.

8.4 BIOPLASTICS SPECTRUM AND ENVIRONMENTAL IMPACT OF BIO-BASED MATERIALS

The Principles for Sustainable Biomaterials gave a valuable structure to assessing the bio-based plastics as of now being marketed. For bio-plastics to be substituted for petroleum-based plastics, they have to meet execution necessities for a scope of items. As of now, bio-based plastics are utilized in strands, films, coatings, bundling, attire, toys, inks, sustenance holders, silverware, and cutlery. Bio-plastics are not broadly utilized in other item classes either due to cost imperatives, execution constraints or absence of information about accessible substitutes that are industrially accessible.

Although bio-based plastics may be more sustainable than petroleum-based plastics in some aspects (see Table 8.1), this analysis found that there are environmental and occupational health and safety hazards in their production. Table 8.2 lists the environmental and occupational health and safety hazards in the production of bio-based plastics. Although advances have been achieved, fully sustainable bio-based plastics with all the highly valued properties of conventional plastics for all types of products are not yet available.

Despite the fact that bio-based plastics might be more supportable than oil-based plastics in certain perspectives (see Table 8.1), this investigation found that there are ecological and word related wellbeing and security risks in their generation. Table 8.2 records the ecological and word related wellbeing and security perils in the generation of bio-based plastics. Despite the fact that progress has been accomplished, completely manageable bio-based plastics with all the very esteemed properties of customary plastics for a wide range of items are not yet accessible.

Table 8.1: Sustainability improvements of bio-based plastics relative to petroleum-based plastics

Bioplastic	Sustainability improvement
polyhydroxyalkanoates	highly biodegradable
polylactic acid	production uses 30–50% less fossil energy and generates 50–70% less CO_2 emissions than PBP; competitive use of water with the best performing PBP, recyclable, compostable at temperatures above 60°C
thermoplastic starch	production requires 68% less energy than its PBP counterpart; lower CO_2 emissions than PBP; biodegradable and compostable
biourethanes	production requires 23% less energy and 36% less GGH, compared with PBP
cellulose and lignin	the biological degradation of lignin is lower than cellulose, compostable
polytrimethylene terephthalate	production requires 26–50% less energy and 44% lower GHG than its PBP counterpart; no chemicals additives are used; biodegradable; potentially recyclable
Corn zein and soy protein	biodegradable and compostable

Source: Ref. 35. GMOs: genetically modified organisms; GHG: greenhouse gases.

Source: https://doi.org/10.1016/j.jclepro.2011.10.003

Table 8.2: Environmental and Occupational Health and Safety Hazards of Bio-based Plastics

Bioplastic	Environmental hazards	Occupational health and safety hazards
Polyhydroxyalkanoates (PHAs)	Feedstock is grown using methods of industrial agricultural production, including GMOs. Data on energy requirements are controversial.	Exposure to pesticides. Physical extraction of PHAs uses pyridine, methanol, hexane or diethyl ether. Chemical digestion uses sodium hypochlorite, methanol and diethyl ether.
Polylactic acid (PLA)	Feedstock is grown using methods of industrial agricultural production, including GMOs. 1-octanol is ecotoxic and organic tin can build up in living organisms.	Exposure to pesticides, sulfuric acid, tin octoate, 1-otanol and urea. Finely pulverized starch can cause powerful explosions.
Thermoplastic Starch (TPS)	Feedstock is grown using methods of industrial agricultural production, including GMOs.	Exposure to pesticides, glycerol and urea. Finely pulverized starch can cause powerful explosions.
Bio-urethanes (BURs)	Feedstock is grown using methods of industrial agricultural production, including GMOs.	Exposure to pesticides, toluene diisocyanate (TDI), methylene diphenyl isocyanate (MDI), tin derivatives.
Cellulose and Lignin	The process has relatively high energy and water requirements. Emissions of pollutants to air and water during kraft process need to be addressed.	Exposure to elevated temperature and pressure. Exposure to disulfide, sodium hydroxide, volatile toxic, flammable and malodorous emissions of sulfur. Exposure to propionic, acetic, sulfuric, nitric acids.
Poly(trimethylene-terephtalate) PTT	Feedstock is grown using methods of industrial agricultural production, including GMOs. Only 37% (by weight) from renewably sourced material GMOs are used in fermentation of glucose to bio-PDO.	Exposure to pesticides, terephthalic acid, dimethyl terephthalate, and methanol. Finely pulverized starch can cause powerful explosions.
Corn zein and soy protein	Feedstock is grown using methods of industrial agricultural production, including GMOs.	Exposure to pesticides, alcohol or volatile solvents, alkaline and acid substances, and formaldehyde as or glutaraldehyde.
Nano-biocomposites (cellulose and lignin)	The process has relatively high energy and water requirements. Emissions of pollutants to air and water during kraft process need to be addressed. Potential toxicity issues of nanoparticles regarding to incineration, composting or recycling are unknown.	Exposure to elevated temperature and pressure. Exposure to disulfide, sodium hydroxide, isocyanates, volatile toxic, flammable and malodorous emissions of sulfur, as well as to nanoparticles.

GMOs: genetically modified organisms, GHG: greenhouse gases.

Source: https://doi.org/10.1016/j.jclepro.2011.10.003

The Bioplastics Spectrums for Occupational Health and Environment (Figs. 8.3 and 8.4) abridge the information accumulated on bio-based plastics as indicated by supportability criteria and give a visual correlation of the bio-based plastics.

The wellbeing and security sway relative investigation (Fig. 8.3), found that PHAs, PLA, and starch (TPS) (shaded as light green in the Spectrum) are linked to the next bio-based materials. Nevertheless, there are some word related hazards in their generation, these risks were viewed as lower than that of the other bio-based materials. They likewise are completely bio-based materials as they don't contain petroleum-based segments.

The natural investigation, outlined in the Bioplastics Spectrum in Fig. 8.4, found that starch, urethanes, PHA, zein, and soy protein (light green) are favored despite the fact that GMOs of obscure dangers and diligent bioaccumulative and poisonous synthetics (PBTs, for example, pesticides might be utilized in feedstock generation. Bio-based feedstocks are commonly developed utilizing techniques for mechanical rural creation and subsequently noteworthy measures of vitality, water, land, GMOs, harmful pesticides and composts are utilized, which exhaust characteristic assets and can contaminate water, air and soil.

Furthermore, there is a worry about utilizing the land for polymer generation that could be utilized to develop crops for human utilization. Advances in biotechnology make it conceivable to deliver plastic specifically in microorganisms or in hereditarily adjusted harvests, for example, corn. The utilization of GMOs in the advancement of bio-based plastics is a matter of concern since their belongings in the earth are not surely known.

Bio-based plastics that are favored from a supportability point of view will use: a feedstock developed without GMOs and without dangerous pesticides:

- will be prepared without GMOs, dangerous synthetic substances or petroleum-based co-polymers,
- will maintain a strategic distance from dangerous added substances or untested nanomaterials
- will address ecological and wellbeing perils amid their production,
- will not affect the food supply,
- will be adaptable in transfer alternatives, which means it could recycle, and on account of landfilling and cremation or (least prescribed choices), it won't produce poisonous outflows

- will be flexible in disposal options, meaning it could compost or recycled, and in the case of landfilling and incineration or (least recommended options), it will not generate toxic emissions,
- will be vitality and water effective underway,
- will evade the age of unsafe side-effects

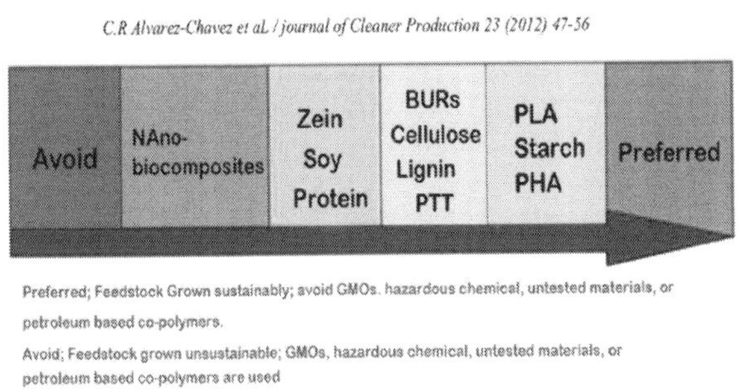

Figure 8.3: The Bioplastics Spectrum. Comparative Occupational Health and Safety impacts of bioplastics.

Source: https://www.packworld.com/sites/default/files/styles/lightbox/public/field/image/Spectrum.jpg?itok=bsgpQTs0

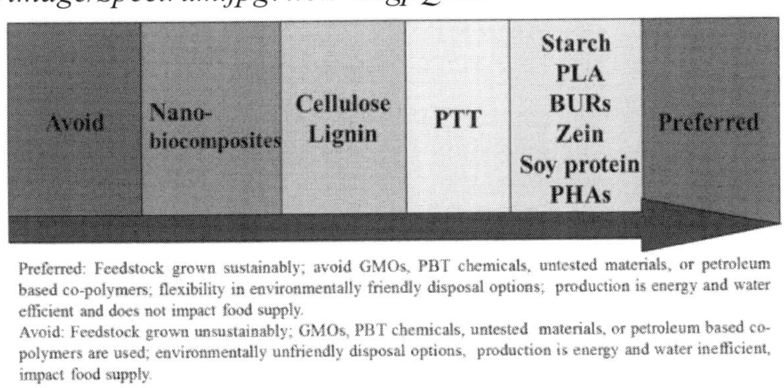

Figure 8.4: The Bioplastics Spectrum.

Source: https://ars.els-cdn.com/content/image/1-s2.0-S095965261100374X-gr6.jpg

8.5 FEEDSTOCK

The decision of feedstock can adversely affect the general manageability of a bio-based item: farming development regularly utilizes noteworthy measures of water, composts, and pesticides. Working conditions on the fields probably wouldn't be consistent with global work guidelines. Moreover, on account of wood, ranger service practices may not be in accordance with supportable ranger service.

Also, bio-based feedstocks supplant fossil-based materials, which are regularly sourced by innovations that are extremely unfavorable to our condition (for example fracking, oil sands or oil slicks through mishaps) and can also impact the society, as well. The privilege bio-based feedstock for an item can be land and asset productive, save money on emanations and be a decent substitute for limited assets.

8.5.1 End-of-Life of bio-based products – issues regarding biodegradability, recyclability etc.

Bio-based items offer elective end-of-life choices contrasted with their regular partners, yet can likewise enter customary transfer courses, for example, reusing or burning. Contingent upon the item's utilization, not all finish of life alternatives bodes well from a natural point of view. Bio-based items can be biodegradable, which can separate them from numerous customary items and can be valuable from various perspectives since new biomass can be secured from it and the asset can be utilized a few times.

In the event that bio-based items go into a procedure of anaerobic absorption after their lifetime, biogas can be created and what remains can be utilized as supplements for horticultural purposes. Notwithstanding, recyclability is likewise an essential issue for maintainability, and handling vitality request just as the transport to fertilizing the soil offices additionally should be thought about when settling on the most reasonable end-of-life alternative.

Besides, not every single biodegradable item gets degraded under similar conditions. Regularly, high temperatures are essential, and in practically all cases biodegradability does not imply that items will debase in the open condition, for example, on the ground or in water, however, require controlled situations with the end goal for this to occur (as in for example modern treating the soil).

As one precedent, biodegradability bodes well inside a closed system, however not really in blended waste streams. For example, for an emergency clinic which has incorporated anaerobic absorption offices that can give vitality specifically to the medical clinic while keeping possibly tainted items on location, the utilization of biodegradable items can give benefits in usefulness and to nature.

8.5.2 Measuring environmental impacts – Life Cycle Assessment (LCA) and Product Carbon Footprint (PCF)

Greenhouse gas (GHG) emissions

The creation and utilization of items and administrations can make diverse gases be transmitted into the climate. As per the Kyoto Protocol, the six most pertinent gases are carbon dioxide (CO_2), methane (CH_4), nitrous oxide (N_2O), hydrofluorocarbons (HFCs), perfluorocarbons (PFCs) and sulfur hexafluoride (SF_6).

They are in charge of retaining and producing radiation at explicit wavelengths inside the range of infrared radiation discharged by the Earth's surface, the climate and mists. This can result in an expansion of worldwide temperatures, normally alluded to as the greenhouse gas effect, with the referenced gases frequently alluded to as GHG.

Life cycle assessment (LCA)

LCA is a worldwide institutionalized technique used to assess an item's effect on the earth over the whole time of its life – from support (that is, from crude material extraction) to its end(that is, transfer or reusing, and so on.).

GHG emanations are one part of an ecological LCA to demonstrate the effects identified with an unnatural weather change of the atmosphere. Other natural effect classes that can be considered are non-sustainable power source use, fossil assets exhaustion, eutrophication, fermentation, and so forth.

LCAs can be useful in an open obtainment procedure to assess the data sources, yields and potential natural effects of buying a specific item for an amazing duration cycle. Life cycle contemplations of items as well as materials are incorporated as a major aspect of some eco-marks, particularly type I and type II eco-names.

Product Carbon Footprint (PCF)

A carbon footprint estimates the complete GHG discharges caused directly and indirectly by an individual, association, occasion or item. A carbon footprint is estimated in huge amounts of carbon dioxide equal (tCO2e) and considers the six Kyoto Protocol GHGs. A PCF is a method for estimating, overseeing and imparting GHGs identified with products and enterprises that is calculated over the length of raw material acquisition to the final reusing or transfer. A carbon footprint is produced dependent on LCA however centers around a solitary issue: dangerous global warming.

Certifications and Labels

Supportability affirmations and names can assist open procurers with defining their prerequisites in relation to bio-based items in their open tenders.

Under the new EU Public Procurement Directives (2014) contracting experts have the likelihood of utilizing marks as a wellspring of data for characterizing specialized details or grant criteria. Names likewise offer a beyond any doubt approach to procurers to check consistency with these necessities, by tolerating the name as one method for verification of consistency with the specialized particulars, for example.

Further subtleties administering the prerequisites for utilizing names in the open acquisition are given in the Directives themselves and in the European Commission's most recent direction on green open acquirement: the "Purchasing Green!" Handbook – the third release was distributed in April 2016.

The below is a (non-exhaustive) list of various labels, certification schemes and benchmarks that might be viewed as when acquiring bio-based products or services:

ISO 14024 Type I Ecolabels specifying bio-based products:

EU Ecolabel ("the Flower"): Lubricants: Lubricants receiving the EU Ecolabel need to prove a minimum content of bio-based carbon between >45% and >70% (depending on the type of lubricant).

Nordic Ecolabel ("Nordic Swan"):

Sanitary products: This product group includes bio-based materials in the form of fluffy pulp and bio-based polymers. The material composition must either include a share of 7% of all polymers from renewable raw materials, a share of 50% of all materials derived from renewable resources or the global warming potential of all the materials cannot be above 2.1 kg CO_2-equivalents/ kilogram of the product.

Disposables for food: A minimum of 90% by weight of the product must be produced from renewable raw materials (except for inorganic fillers which can make up to 20% of the product).

Blue Angel ("Blauer Engel"):

Office materials (writing utensils and stamps): A very recent revision of the criteria for these products has included the provision that they need to be produced either from 60% renewable raw materials or from 80% recycled materials.

8.6 RELATIONSHIP BETWEEN BIOPLASTICS AND SUSTAINABILITY

Role of Biomaterials in Sustainability 5

However, long lead times, and high research and development costs are burdensome on the parties associated with their development, including farms.

Bio based chemicals, and their raw materials, provide more predictable development requirements and markets compared to pharmaceuticals. The potential bio based chemicals with sizeable markets include 1,3-Propanediol which is a building block of biopolymers that is made from maize syrup by modified E. Coli bacteria. The world market for this material is estimated to be 0.23 million tons in 2020 (Carole, Pellegrino, & Paster, 2004). Specialty chemicals functioning as adhesives, solvents, and surfactants are crucial in detergents, cosmetics, and manufacturing processes. Surfactants, still mainly petroleum based, are increasingly fabricated from renewable materials such as tropical vegetable oils -with a production volume of more than 2 million tons (Turley. 2008). Longer chained-oils from rapeseed and sunflower are desirable for use in surfactants, as well as lubricants, adhesives, solvents, and biopolymers. Solvents used in the manufacturing of pharmaceuticals, paints, and inks are made from bio based products such as ethyl lactate (Carole, Pellegrino, & Paster, 2004). Alcohols and fatty acids, obtained through fermentation of carbohydrates from cereals, potato, and sugar beets, are employed in the production of lactate esters including ethyl lactate. Biolubricants offer another innovative area for agriculture and industry for lubricants and hydraulic fluids. In particular, bio based hydraulic fluids and inks can exhibit high performance characteristics (Nowicki, Banse, Bolck, Bos, & Scott, 2008; Lazerri, 2009).

Relationship Between Bioplastics and Sustainability

An effort to establish a strong ties between biodegradable plastics (bioplastics) and sustainability directs us to recent engineering advances.

An emerging area of engineerin Role of Biomaterials in Sustainability 5

However, long lead times, and high research and development costs are burdensome on the parties associated with their development, including farms. Bio based chemicals, and their raw materials, provide more predictable development requirements and markets compared to pharmaceuticals. The potential bio based chemicals with sizeable markets include 1,3-Propanediol which is a building block of biopolymers that is made from maize syrup by modified E. Coli bacteria. The world market for this material is estimated to be 0.23 million tons in 2020 (Carole, Pellegrino, & Paster, 2004). Specialty chemicals functioning as adhesives, solvents, and surfactants are crucial in detergents, cosmetics, and manufacturing processes. Surfactants, still mainly petroleum based, are increasingly fabricated from renewable materials such as tropical vegetable oils -with a production volume of more than 2 million tons (Turley. 2008). Longer chained-oils from rapeseed and sunflower are desirable for use in surfactants, as well as lubricants, adhesives, solvents, and biopolymers. Solvents used in the manufacturing of pharmaceuticals, paints, and inks are made from bio based products such as ethyl lactate (Carole, Pellegrino, & Paster, 2004). Alcohols and fatty acids, obtained through fermentation of carbohydrates from cereals, potato, and sugar beets, are employed in the production of lactate esters including ethyl lactate. Biolubricants offer another innovative area for agriculture and industry for lubricants and hydraulic fluids. In particular, bio based hydraulic fluids and inks can exhibit high performance characteristics (Nowicki, Banse, Bolck, Bos, & Scott, 2008; Lazerri, 2009).

Relationship Between Bioplastics and Sustainability

An effort to establish a strong ties between biodegradable plastics (bioplastics) and sustainability directs us to recent engineering advances. An emerging area of engineerinA push to set up – solid ties between biodegradable plastics (bioplastics) and supportability guides us to ongoing building progresses. A rising region of designing materials is self-dissolving plastics materials which are manageable as well as draw motivation from organic frameworks which show comparable qualities.

Bioplastics guarantee a wide scope of conceivable outcomes. Starch-based plastic have been accessible for a couple of decades and can be gotten from cassava, maize and wheat. Starch properties, for example, the amylose/amylopectin proportion and starch granule estimate are essential for success. Amylose ethers are biodegradable elective feedstocks to polyethylene and polystyrene.

Polyesters, for example, polylactic corrosive and polyhydroxyalkanoate are aggressive with oil-based polymers like PET (polyethylene terephthalate) and are produced using starch or sugar by means of maturation. The principle crude material for polylactic corrosive is glucose syrup produced using maize, stick, potato or wheat. Starch-based bioplastics are normally used in packing materials, kitchenware, vehicle insides, and diapers.

Regular strands can be connected to composite materials utilizing cellulosic crude materials from wood and straw or yields, for example, kenaf, sisal, jute and hemp. Eucalyptus may supplant synthetics like rayon. Composite materials with cellulosics offer decreased weight, improved wellbeing, and great acoustics. Regular filaments can likewise be utilized to fortify manufactured materials, as opposed to supplanting them.

Corn is the crude material that a large portion of the bioplastic makers depend on. Nature Works situated in Minnetonka Minnesota created one of the primary plant-based plastics. The tar created was called Ingeo and used to make attire, diapers, and sustenance packing materials. The organization is a backup of agribusiness mammoth Cargill and Teijin of Japan and evaluated in 2008 as a future interest of 50 billion pounds of bioplastics every year within two to five years. Certain issues make biodegradable plastics doable – the economy of making them, their relative exhibitions versus regular materials, and their adequacy in compostability. This investigation won't incorporate the execution examinations among ordinary and bio-based materials since numerous fruitful bio-based applications exist without limitations. In any case, extra data on the effect of bio-based materials to out condition are given.

The fundamental US technique for bundling waste transfer is landfill, trailed by reusing, cremation, and treating the soil. Utilization of landfills is unfortunate because of field space requirements and age of contaminants including ozone-harming substances.

Reuse, reusing, and fertilizing the soil is expected to improve the effect of the waste transfer to our surroundings. A large portion of the basic packing materials, for example, aluminum, glass, paper and wood are recouped effectively by reusing except if they are debased with natural substances including nourishment. As new biopolymers are created and progressively used in sustenance, pharmaceutical and customer products, fertilizing the soil could wind up one of the fundamental techniques for transfer of packing waste.

Biopolymers, for example, polylactide and polyhydroxy butyrate are getting to be accessible for use in different packaging applications. The reason for these new biomaterials is that they are obtained from sustainable assets and that they can be degraded in soil and manure effectively. Since arranging and cleaning needs invalidate the upside of lesser vitality prerequisite of reusing, the primary way is to discard them by fertilizing the soil.

The fundamental drivers for these biopolymer-based packaging have been green advertising and pseudo-ecological awareness identified with high fuel costs. Packaging compostability could be an option for the transfer of bio-based materials as long as a society all in all is eager to formally allow such an eco-friendly pourpose.

8.7 BIOPLASTICS OFFER A MORE SUSTAINABLE FUTURE

As the world's biggest maker of plastics produced using plants, NatureWorks LLC, a US headquartered company that manufactures bioplastics wish to elucidate John Vidal's cases about bioplastics (Sustainable bioplastic can harm the earth, April 26). While bioplastics are still on an adventure to full manageability, the company believes that bioplastics are the safer alternative in contrast to oil-based plastics.

The advantages of bioplastic start in the manufacturing process. Peer-reviewed studies have found that making Ingeo - known generically as polylactic acid or PLA - generates less than half the greenhouse emissions of making oil-based plastics. New innovations in the manufacturing process mean Ingeo production will soon produce 75% lesser greenhouse emissions than making Petro-plastics.

The ecological advantages proceed through the term of the plastic. In contrast to customary plastics, PLA is vastly recyclable: a PLA jug can be made into another container over and over, though oil-based contenders are ordinarily "down-cycled" into results of lessening esteem and eventually bound for the landfill. Ingeo is additionally affirmed as completely compostable in explicit conditions. The Association for Organics Recycling (earlier the UK Composting Association) underpins the expanded utilization of biodegradable material as a method for lessening the 4.6m huge amounts of packaging discarded every year in the UK.

Affordable food, solid vitality and a sound situation are worldwide issues that can't be tackled without central changes the same way we deal with limited natural resources.. The company trust that plants, as opposed to petroleum derivatives, are a decent spot to begin.

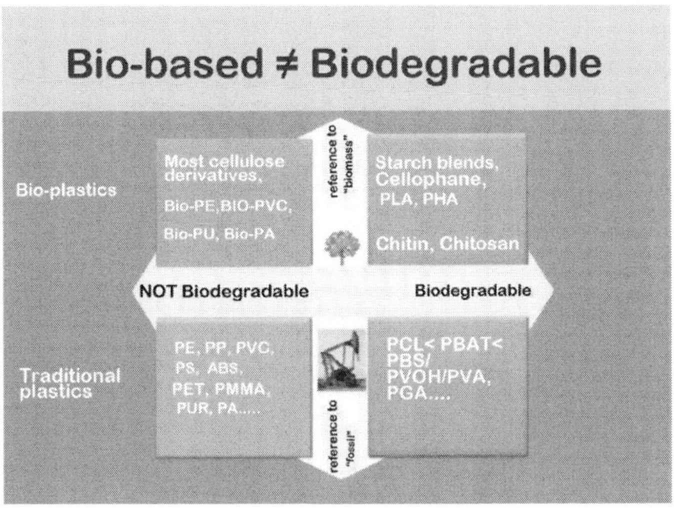

Figure 8.5: Are bio-based plastics really sustainable?.

Source: https://www.slideshare.net/KeepAsiaBeautiful/are-biobased-plastics-really-sustainable-74864325

8.8 CREATING SUSTAINABLE PLASTICS

Shoppers need to like the items they buy. Some portion of that is maintainability—essentially, they need to purchase something considered "green" since it makes them feel like they are accomplishing something useful for the Earth. This section intends to demonstrate how item architects and makers can best satisfy the developing need for increasingly maintainable items without yielding quality or cost.

There are numerous approaches to make plastic increasingly practical – from utilizing progressively sustainable substance and recovering and reusing material, to decreasing vitality required for assembling and returning material to nature toward an amazing finish.

In addition, it is conceivable to make increasingly practical plastics with one of a kind physical exhibition and stylish characteristics. In this piece, the manners in which organizations can make progressively feasible plastic items that are greener, yet in addition, perform better is explained briefly.

8.8.1 The Move toward Sustainability

Green is the new standard. In spite of the logical circumstances and the logical results of numerous ecological issues remain bantered by a few, what is obvious is a developing shopper enthusiasm for earth neighborly organizations and items. The interest in maintainable items is developing, and organizations are profiting by the rising interest for earth well-disposed items.

In 2012, the Accenture counseling organization directed a study of 250 senior administrators in the U.K., U.S., Japan, Germany, France, China, Brazil, and India. They found that by far most said manageability was essential to the future development of their organizations. Of those studied, 62% said their ventures were inspired by client desire for reasonable items and were seen as a chance to drive development.

"Fortunately, organizations presently methodically consider manageability to be being fundamental to their future development and center to their business," said Bruno Berthon, overseeing chief of supportability administrations at Accenture. "Manageability has broken free from the domain of administrative weight and notoriety the board and is currently ascending into a temperate hover of business opportunity and speculation development." More than 33% of the examination's respondents said that they couldn't stay aware of purchaser interest in supportable items.

A recent report led by the Association for Enterprise Opportunity titled, "The Big Green Opportunity,"2 overviewed more than 1,300 private ventures in the U.S. Furthermore, found that as per respondents, the greener an item, the more probable its deals will increment, with 75% revealing an expansion in green business from 2008-2011 even at a 19% value premium contrasted with customary products.

79% of the private companies studied said that offering green items or administrations gave their business an upper hand. That upper hand is being filled by the development in buyer interest for progressively maintainable items. As indicated by the creator of the examination, Denise Hamsler Hamler, "green sections today are becoming far quicker than their general ventures."

8.8.2 How Consumers Understand Sustainability

The development behind sustainability is consumer driven. Shoppers need to purchase something "green." Consumers need to think about the materials utilized in an item, how and where it is made, how far it voyages and how it is bundled, yet they may have changing criteria for what makes an item progressively economical.

Effective organizations are separating their items by plainly expressing the ecological advantages of their item, regardless of whether it be utilizing progressively sustainable materials, utilizing recovered or reused materials, or materials that can be recycled once their use is over. There is a developing assortment of increasingly economical plastics. As the interest keeps on rising, so will the use of these materials in applications.

Manageable feedstocks, for example, plant fibers, wood, and starches are progressively being utilized in plastics; these elective feedstocks are promptly accessible and moderate are available easily. These sources can be utilized to supplant oil-based feedstocks. Truth be told, huge numbers of these bioplastics have benefitted that conventional plastics don't. No single definition for manageability exists. There are various attributes that an item can have—both free of each other and together—that will decide an item's maintainability.

To make things somewhat less difficult, few definitions here you are listed to remember as you assess your own organization's supportability objectives. Bioplastics are a group of materials with contrasting properties and applications. A material is characterized as a bioplastic on the off chance that it is either biobased, biodegradable or highlights the two properties. Bio-based materials are produced using natural (carbon-based) materials that contain in entire or part biogenic (organic) carbon – supplanting petroleum/ fossil carbon with bio/inexhaustible carbon.

In the United States, the USDA Bio preferred Program ensures materials as bio-based. Items made with ensured bio-based materials can get particular obtaining from Federal acquirement offices. The level of sustainable material important to meet this prerequisite fluctuates as per the item.

- *Bio composite materials* join biomaterials like wood, starch, flax, jute, hemp, and other comparative materials with conventional plastics. These materials are utilized to upgrade physical attributes while decreasing the measure of non-sustainable oil-based plastic.
- *Biodegradable* is the capacity of a material to experience biodegradation, a compound procedure amid which microbes convert materials into basic substancesusing natural conditions.
- *Compostable* items are characterized by ASTM International

gauges 6400 and 6868. These models necessitate that the material biodegrades in a specific timeframe and leaves no poisonous buildup in the soil.

The ASTM compostability benchmarks are commonly expected to apply to packing, not solid items. Note that utilizing ASTM tried compostable materials may not imply that your item will satisfy these guidelines. This will rely upon the size, volume and thickness of the item.

Plastic reusing is the way toward recouping scrap or waste plastic and reprocessing the material into helpful items, at times totally unique. For example, this could mean softening down soda pop containers and afterward shaping them as plastic seats and tables. The resuing thus, depends on innovation and the gathering of utilized plastics.

The carbon footprint of items depicts the total of ozone-depleting substance gases amassed amid the full life cycle of an item in a predefined application. Ozone harming substances trap heat in the environment and are viewed as a critical reason for worldwide environmental change.

Toxicity ought to likewise be considered for example, synthetic compounds like bisphenol A and phthalates used in some oil-based plastics harm hormones Expanding quantities of buyers are looking for plastic materials that are free from these synthetic compounds.

The mind-boggling framework of variables adding to supportability makes it critical to characterize your manageability goals. Is your objective to make an item with progressively inexhaustible materials? Is it to utilize progressively recovered and reused materials? Is it to most likely return these materials back to nature? Or then again would you say you are looking for a blend of ecological advantages? As should be obvious, the idea of maintainability is more perplexing than you may have thought. Thus, it's vital to characterize the maintainability objectives for your item and pick a bioplastics accomplice that can give materials to meet your particular needs.

Obviously expressing what your item is produced using, how and where it's made, and why it's better will help separate your item. At the point when clients know why an item is increasingly manageable, they'll get behind it and like their decision.

8.8.3 The Benefits of Sustainable Plastics

While 78% percent of purchasers state they need to purchase progressively feasible items, just 43% will pay more. Customers would prefer not to forfeit quality or cost for the things they purchase. Envision a customer item produced using the "immaculate plastic," one that meets pretty much every maintainability standard. The downside, in any case, is that this ideal plastic item costs twice that of its conventional plastic partner. Purchasers are far less inclined to purchase this ideal plastic item, basically in view of the cost. On the off chance that an item does not speak to shoppers, the ecological advantages will never be figured it out.

Take an electric vehicle, for instance. While electric autos can significantly affect ehaust the cost has been a deterrant. At the point when a great many gas-controlled autos run more effectively and produce lesser outflows, the amassed gradual improvement has a far more noteworthy positive effect on nature. Also, improving the supportability of the plastics items that we utilize ordinary can have a noteworthy environmental effect.

Supplanting oil-based plastics with sustainable, recovered or reused materials can be the least complex approach to help the natural impression of items. These materials can decrease the measure of non-sustainable feedstocks while expanding the physical and stylish qualities of an item.

Bio composites produced using mechanical results like recovered wood strands and reused plastic can urge makers to lessen landfill waste. Utilizing byproducts can build crop esteems for ranchers without contending with food crops. Utilizing compostable materials can urge buyers to restore these items to nature when their valuable life has finished, further lessening landfill waste, and now and again, diminish the expense of arranging these materials.

Various physical advantages result from the consideration of biomaterials in plastics also, for example, expanded quality, solidness and toughness. Moreover, items produced using bio-based materials essentially look and feel increasingly regular. Utilizing starch-based plastic can upgrade items with a gentler touch. These items can likewise be effectively scented to give further item separation. Wood and characteristic filaments can give a characteristic look that is more tastefully satisfying.

8.8.4 Meeting Sustainability Goals by Choosing the Right Bio-plastic Partner

Picking the best material for your item can be a troublesome errand. That is the reason for working with the privilege bioplastics organization. Numerous bioplastics organizations can just offer one sort of material that might possibly address your issues. An organization with a wide scope of bio composites, bio-based, and compostable materials can give you choices best coordinated to your supportability objectives.

The ability to tweak the definitions of these materials to meet your particular undertaking necessity is additionally an essential factor. Your bioplastics accomplice ought to have the assets and skill to rapidly build up these materials to meet the physical prerequisites of your item.

It's vital to work with plastics experts that have a profound comprehension of how plastic items are made. Your bioplastics accomplice ought to have assets and specialized mastery in plastic preparing to give help with instrument making and assembling. This skill can be especially useful in discovering toolmakers and assembling accomplices to create your undertaking.

Picking a bioplastics organization that is focused on serving you all through item improvement and assembling is urgent to the accomplishment of your task.

8.8.5 The Future of Sustainable Plastics

Today bioplastics are under 2% of the whole plastics showcase however buyers and retailers are as of now requesting progressively practical materials and this interest will just keep on expanding. As indicated by late examinations, interest for bioplastics is anticipated to fourfold by 2018. On the off chance that makers expect to remain aggressive and fulfill this need, they should make increasingly manageable items that don't forfeit expense or quality.

We are at the beginning of another time of plastics, one in which plastic items are made with a wide scope of inexhaustible, recovered or compostable materials. In our lifetime we may see conventional oil-based plastics as a dinosaur, much the same as the steam train, or utilizing coal heaters to warm our homes. Future ages may chuckle at utilizing valuable non-sustainable feedstocks to make expendable ordinary items.

By acquainting new creative materials with makers and buyers we can offer better decisions for the items they need and utilize regularly. These materials will outflank customary oil-based plastic. These steady enhancements won't unravel the majority of our ecological difficulties, however, they can have a critical positive effect.

The right bioplastics partner can enable you to fulfill developing need for increasingly economical plastic items, providing materials that are not only greener but on the other hand, are more secure, perform better and last more.

8.9 FACING THE CONTRADICTIONS OF PLASTICS

A legitimate evaluation of plastics as being helpful, essential materials additionally requires conceding that plastic items shapes, frames, arrangements, uses, and material characteristics are fairly baffling or conflicting. Plastic items in some cases take after articles from nature, albeit here and there an outsider, even sci-fi type of nature. Be that as it may, regularly they are not considered as resembling whatever is normal; all things considered, plastic syntheses are made by scientific experts and are baffling to those uneducated in natural science.

The term 'plastics' is itself lacking and deceiving in that it alludes to a wide scope of materials. A few plastics are rubbery, and some don't dissolve when warmed; some are reinforced with glass fiber to wind up composites, while others are utilized as straightforward movies or froths. Certain plastic items are feeble, modest and dispensable; others are solid and tough – yet all are named just as plastic.

Different logical inconsistencies beneath identify with plastics place in our economy and in our bigger regular habitat:

- Plastics are modest.... however, their properties can be custom made for each high-esteem, designing purposes
- Plastics regularly have moderately straightforward concoction structures...yet they can be incredibly impervious to procedures of natural decay, ensuring their long-haul ingenuity in nature
- Plastic items are lightweight.... yet huge amount of them have been utilized and discarded.
- Plastics are produced using high vitality compound feedstock, getting to be basically "solidified fuel" ...yet plastic items and their inalienable energy are usually treated as waste for landfills, frequently after exceptionally concise lifetimes of utilization (in contrast to fluid or vaporous fuels, whose energy content is converted directly into heat or motion)
- Most waste plastics can be reprocessed at moderately low temperatures and energies...yet their relative ease implies their gathering, division and reusing is frequently not cost-focused with the creation of new, virgin plastic

- Different plastics are considered by the people in general as being fundamentally the same as and their disparities are regularly difficult to recognize by eye…yet their custom fitted sciences, mechanical properties, and added substances content frequently attach every plastic to a particular use (which additionally makes them harder to reuse and reuse in numerous applications).

Along these lines, despite the fact that plastic items are adroitly planned and created for even the least difficult utilization… most are still normally bound to end up (entirely solid) garbage.

Bio-based polymers right currently incorporate the undeniably prominent corn ethanol-based polylactic corrosive (PLA). In any case, the bio-based field may before long turn out to be commanded by traditional, surely knew polymers, for example, polyethylene and polypropylene, that depend on feedstock produced using plant assets, for example, Brazilian sugarcane rather than non-renewable energy sources.

For instance, Braskem SA in 2010 opened a plant in Brazil to make 400 million lb./yr of sugarcane-based polyethylene. The organization has additionally devoted $100 million for a 66-million lb./year bio-based polypropylene plant slated to start the activity by 2013. Despite the fact that these materials will at first be evaluated at a premium contrasted and petroleum derivative based PE and PP, they remain as signs that the business is without a doubt ready to create bio-based choices for plastics.

8.10 PLASTICS AT PLAY IN CONSUMER LIFESTYLES

Another plastic logical inconsistency concerns customers both in developed nations and, progressively in developing nations. Customers have turned out to be extraordinarily reliant on plastic materials, even without understanding a considerable lot of the subtleties of their creation or how they are produced.

There is something practically awesome about plastic items – their shapes and hues, their surfaces and smooth structures. In the energizing beginning of plastics, individuals may have even been enticed to reference sci-fi creator Arthur C. Clarke's "Third Law" from 1962: "Any adequately trend-setting innovation is indistinct from enchantment".

Be that as it may, presently, after never truly accomplishing the dramatic intrigue of room flight, PCs or the wireless, plastics are never again new. The most well-known prevalent polymers have outlasted their innovators, and any feeling of plastics enchantment or interest for purchasers is lost behind the triviality of the numerous items they are utilized in. Or maybe, a most normal inquiry caught today would be: "Do we truly require plastics?"

Accordingly, numbness, lack of interest, antagonistic vibe or uncertainty may be better terms of depicting purchaser's perspective on plastics today. These perspectives have brought about the simple belittling of specific types of plastics, with total negligence about other potential issues related with the manners in which we utilize all materials in industrialized society – metals, woods, and minerals, just like plastics.

People have expended a greater amount of every one of these materials in the previous 50 years than in every single earlier year joined – at expanding appraised. Also, most by far of modern materials are not founded on normally recharging assets, not at all like in 1990, when over 40% of all materials utilized in the United States were situated in ranger service, agribusiness or other sustainable businesses. These patterns, combined with inconceivable populace development, have certainly focused on the earth.

During the 1990s, wellbeing concerns emerged identified with the phthalate substance plasticizing added substances used to diminish PVC. Normally utilized phthalate plasticizers are not artificially fixing to the polymer spine, so they tend to relocate to the outside of the item. Their impact on youngsters was of specific concern, and kids were the main impetus of the discussion since they will in general spot adaptable vinyl toys and items into their mouths.

Following many studies about how phthalates upset the endocrine arrangement of people, and counter investigations saying the inverse (frequently subsidized by an industry that depends on utilizing phthalates for PVC) guidelines started to be passed that confined the utilization of normal phthalates. Aside from inside and out bans in some western nations, the US Consumer Product Safety Commission in 2009 limited to extremely low focuses various regular phthalates in items utilized by kids under twelve years of age. Accordingly, added substance providers have made lines of plasticizer options, some of which are bio-based or which cling to the PVC polymer network.

Maybe there are exercises to be gained from this history of PVC, as new bio-based plastics are presented. Some gum providers are as of now considering inquiries like the accompanying before the vast scale advertising of their non-renewable energy source based gums or their bio-resins:

- What mixes are discharged amid the creation of the crude polymers, and amid their exacerbating? Is it accurate to say that they are perilous to laborers? Would any administrative office think about them so?

- What added substances are aggravated into the plastic? Are any disputable? Have they been tried and affirmed to be non-toxic in each circumstance in which a client may experience them? It is safe to say that they are firmly bound in the last plastic organization, or would they be able to be separated from the plastic part amid use?
- Can a reusing framework conceivably be made for the material, in the event that it doesn't as of now exist ?

REFERENCES

1. Edwards, Sally (2011). Sustainability of bio-based plastics: general comparative analysis and recommendations for improvement. Journal of Cleaner Production. www.elsevier.com/ locate/jclepro
2. Creating Sustainable Plastics Without Sacrificing Quality or Cost. Green Dot
3. Sustainability of bio-based products. Factsheet No.2. InnProBio (Forum for Bio-based Innovation in Public Procurement)
4. Connecting Bio-based forces for a Sustainable World. Biovoices

INDEX

U

V

W